Before You Turn the Page,
Take the **TCAP Online Practice Test**

For Students & Parents

1. Scan the QR Code or Visit:
   lumoslearning.com/book

2. Enter the Access Code: **KWS3**

3. Start the test

For Educators

1. Scan the QR Code or Visit: **lumoslearning.com/book**
2. Enter the Access Code: **TLZ4**
3. Get Started

Tennessee Comprehensive Assessment Program Test Prep: 8th Grade Math Practice Workbook and Full-length Online Assessments: TCAP Study Guide

Contributing Author - Nicole Fernandez
Contributing Author - Nancy Chang
Contributing Author - Greg Applegate
Executive Producer - Mukunda Krishnaswamy
Program Director - Anirudh Agarwal
Designer and Illustrator - Snehal Sharan

Updated in July 2026

ISBN 13: 979-8897131273

Printed in the United States of America

FOR SCHOOL EDITION AND PERMISSIONS, CONTACT US

LUMOS INFORMATION SERVICES, LLC

PO Box 1575, Piscataway, NJ 08855-1575
www.LumosLearning.com

Email: support@lumoslearning.com
Tel: (732) 384-0146
Fax: (866) 283-6471

Lumos Learning
Step Up Your Skills

How to Use the Lumos Program?

Congratulations on choosing the Lumos TCAP Program, a blended program that combines online practice with paper-and-pen skill building.

How to Use the Lumos TCAP Program:

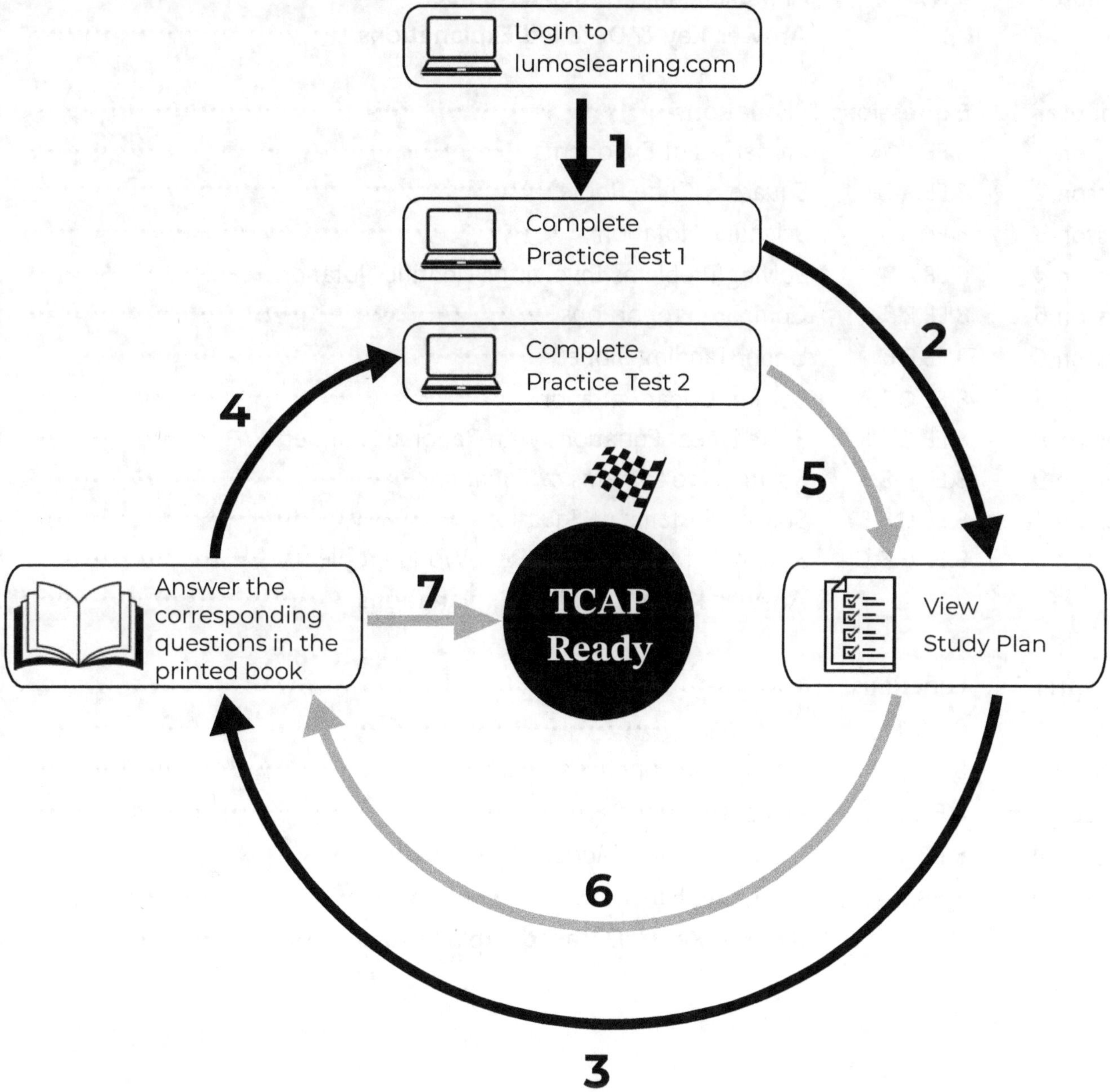

Table of Contents

Date of Completion:_____________ Score:_____________

Chapter 1: The Number System

Lesson 1: Rational vs Irrational Numbers

1. Which of the following is an integer?

   Ⓐ -3
   Ⓑ $\frac{1}{4}$
   Ⓒ -12.5
   Ⓓ 0.454545...

2. Which of the following statements is true?

   Ⓐ Every rational number is an integer.
   Ⓑ Every whole number is a rational number.
   Ⓒ Every irrational number is a natural number.
   Ⓓ Every rational number is a whole number.

3. Which of the following accurately describes the square root of 10?

   Ⓐ It is rational.
   Ⓑ It is irrational.
   Ⓒ It is an integer.
   Ⓓ It is a whole number.

4. Complete the following statement: Pi is ________ .

   Ⓐ both real and rational
   Ⓑ real but not rational
   Ⓒ rational but not real
   Ⓓ neither real nor rational

5. Complete the following statement: $\sqrt{7}$ is _________.

   Ⓐ both a real and a rational number
   Ⓑ a real number, but not rational
   Ⓒ a rational number, but not a real number
   Ⓓ neither a real nor a rational number

6. The number 57 belongs to which of the following set(s) of numbers?

Ⓐ N only
Ⓑ N, W, and Z only
Ⓒ N, W, Z, and Q only
Ⓓ All of the following: N, W, Z, Q, and R

7. From the following set: {$-\sqrt{5.7}$, -9, 0, 5.25, 3i, $\sqrt{16}$}
Select the answer choice that shows the elements which are Natural numbers.

Ⓐ $-\sqrt{5.7}$, -9, 0, 5.25, 3i, $\sqrt{16}$
Ⓑ $-\sqrt{5.7}$, -9, 0, 5.25, 3i
Ⓒ 3i
Ⓓ Positive square root of 16

8. From the following set: {$-\sqrt{5.7}$, -9, 0, 5.25, 3i, $\sqrt{16}$}
Select the answer choice that shows the elements that are Rational numbers.

Ⓐ $-\sqrt{5.7}$, -9, 0, 5.25, 3i, $\sqrt{16}$
Ⓑ -9, 0, 5.25, $\sqrt{16}$
Ⓒ 3i
Ⓓ $-\sqrt{5.7}$

9. Which of the numbers below is irrational?

Ⓐ $\sqrt{169}$
Ⓑ $\sqrt{4}$
Ⓒ $\sqrt{16}$
Ⓓ $\sqrt{3}$

10. Write the repeating rational number 0.1515... as a fraction.

Ⓐ $\frac{85}{100}$

Ⓑ $\frac{15}{75}$

Ⓒ $\frac{15}{99}$

Ⓓ $\frac{25}{50}$

11. Write the repeating rational number .112112... as a fraction.

Ⓐ $\frac{112}{100}$

Ⓑ $\frac{112}{99}$

Ⓒ $\frac{112}{999}$

Ⓓ $\frac{111}{999}$

12. Which of the following is true of the square root of 2?

Ⓐ It is both real and rational.
Ⓑ It is real but not rational.
Ⓒ It is rational but not real.
Ⓓ It is neither real nor rational.

13. Which of the following sets includes the square root of -25?

Ⓐ R, Z, W, N, and Q
Ⓑ R, W, Q
Ⓒ Z, N
Ⓓ None of the above.

14. Complete the following statement:
The number 6.25 belongs to ____________.

Ⓐ R, Q, Z, W, and N
Ⓑ R and Q
Ⓒ R and N
Ⓓ Q and Z

15. **Complete the sentence:**
Irrational numbers may always be written as _________.

Ⓐ fractions
Ⓑ fractions and as decimals
Ⓒ decimals but not as fractions
Ⓓ neither decimals or fractions

16. **Which of the following are rational numbers?**

Instruction : Mark all the correct options. More than one option may be correct.

Ⓐ $\frac{5}{7}$
Ⓑ $\sqrt{10}$
Ⓒ $\sqrt{25}$
Ⓓ π

17. **Mark whether each number is rational or irrational.**

	Rational	Irrational
$\sqrt{2}$		
$\frac{1}{3}$		
0.575		
$\frac{\sqrt{12}}{4}$		

18. **Identify the irrational number and circle it.**

Ⓐ $\frac{5}{7}$
Ⓑ 0.1
Ⓒ $\sqrt{10}$

CHAPTER 1 → Lesson 2: Approximating Irrational Numbers

1. Between which two whole numbers does $\sqrt{5}$ lie on the number line?

Ⓐ 1 and 2
Ⓑ 2 and 3
Ⓒ 3 and 4
Ⓓ 4 and 5

2. Between which pairs of rational numbers does $\sqrt{5}$ lie on the number line?

Ⓐ 2.0 and 2.1
Ⓑ 2.1 and 2.2
Ⓒ 2.2 and 2.3
Ⓓ 2.3 and 2.4

3. Order the following numbers on a number line (least to greatest).

Ⓐ 1.8, 1.35, 2.5, $\sqrt{5}$
Ⓑ 1.35, $\sqrt{5}$, 1.8, 2.5
Ⓒ 1.35, 1.8, $\sqrt{5}$, 2.5
Ⓓ 1.35, 1.8, 2.5, $\sqrt{5}$

4. If you fill in the _____ in each of the following choices with $\sqrt{7}$, which displays the correct ordering from least to greatest?

Ⓐ ___, 2.5, 2.63, 2.65
Ⓑ 2.5, ___, 2.63, 2.65
Ⓒ 2.5, 2.63, ___, 2.65
Ⓓ 2.5, 2.63, 2.65, ___

5. Which of the following numbers has the least value?

Ⓐ $\sqrt{(0.6561)}$
Ⓑ 0.8
Ⓒ 0.8...
Ⓓ 0.8884

6. **Choose the correct order (least to greatest) for the following real numbers.**

Ⓐ $\sqrt{5}, 4\frac{1}{2}, 4.75, 2\sqrt{10}$

Ⓑ $4\frac{1}{2}, \sqrt{5}, 2\sqrt{10}, 4.75$

Ⓒ $4\frac{1}{2}, 4.75, \sqrt{5}, 2\sqrt{10}$

Ⓓ $\sqrt{5}, 2\sqrt{10}, 4\frac{1}{2}, 4.75$

7. **Which of the following numbers has the greatest value?**

Ⓐ 0.4...
Ⓑ 0.444
Ⓒ $\sqrt{0.4}$
Ⓓ 0.45

8. **Which of the following is ordered correctly from least to greatest?**

Ⓐ $\sqrt{0.9}$, 0.9, 0.999, 0.9...
Ⓑ 0.9, $\sqrt{0.9}$, 0.999, 0.9...
Ⓒ 0.9, 0.9..., $\sqrt{0.9}$, 0.999
Ⓓ 0.9, 0.9..., 0.999, $\sqrt{0.9}$

9. **Write the following numbers from least to greatest.**

Ⓐ $\sqrt{2}, \pi, 3\frac{7}{8}, \frac{32}{8}$

Ⓑ $\pi, \sqrt{2}, 3\frac{7}{8}, \frac{32}{8}$

Ⓒ $3\frac{7}{8}, \pi, \sqrt{2}, \frac{32}{8}$

Ⓓ $\frac{32}{8}, 3\frac{7}{8}, \pi, \sqrt{2}$

10. If you were to arrange the following numbers on the number line from least to greatest, which one would be last?

Ⓐ 3.6

Ⓑ $3\frac{7}{12}$

Ⓒ $\sqrt{12}$

Ⓓ $3\frac{9}{10}$

11. Between which of these pairs of rational numbers does $\sqrt{24}$ lie on the number line?

Ⓐ 4.79 and $4\frac{7}{8}$

Ⓑ $4\frac{7}{8}$ and 5.0

Ⓒ 4.95 and 5.0

Ⓓ 4.75 and 4.79

12. Between which two integers does $\sqrt{2}$ lie on the number line?

Ⓐ 0 and 1
Ⓑ 1 and 2
Ⓒ 2 and 3
Ⓓ 3 and 4

13. Between which pair of rational numbers does $\sqrt{2}$ lie on the number line?

Ⓐ 1.40 and 1.41
Ⓑ 1.41 and 1.42
Ⓒ 1.42 and 1.43
Ⓓ 1.43 and 1.44

14. Between which pair of consecutive integers on the number line does $\sqrt{3}$ lie?

Ⓐ 1 and 2
Ⓑ 2 and 3
Ⓒ 3 and 4
Ⓓ 4 and 5

15. Between which of the following pairs of rational numbers on the number line does $\sqrt{3}$ lie?

Ⓐ 1.70 and 1.71
Ⓑ 1.71 and 1.72
Ⓒ 1.72 and 1.73
Ⓓ 1.73 and 1.74

End of The Number System

ANSWER KEY AND DETAILED EXPLANATION

Chapter 1: The Number System

Lesson 1: Rational vs. Irrational Numbers

Question No.	Answer	Detailed Explanation
1	A	An integer belongs to the set containing the counting numbers, their additive inverses, and zero. Therefore, (-3) is an integer.
2	B	Rational numbers are the set of numbers that can be expressed as the quotient of two integers in which the denominator is not zero. All whole numbers can be expressed in this manner; so every whole number is a rational number.
3	B	$\sqrt{10}$ cannot be expressed as the ratio of two integers p and q and is therefore irrational.
4	B	Pi is the ratio of a circle's circumference to its diameter. It is therefore a real number. Pi cannot be expressed as the ratio of two integers, so it is irrational.
5	B	$\sqrt{7}$ cannot be expressed as the ratio of two integers and is therefore irrational. The irrationals are a subset of the real numbers.
6	D	The number 57 meets the requirements of each of the following sets of numbers: N (natural numbers), W (whole numbers), Z (integers), Q (rational numbers), and R (real numbers).
7	D	By definition, the natural numbers, N, are the set of counting numbers. Some mathematicians also include zero in this set. Since $\sqrt{16}$ = +4 or -4 and +4 is a counting number, it is included in N. None of the choices offered 0 as an option; so, in this case, it is a mute point.
8	B	3i is an imaginary number and therefore not rational and - $\sqrt{5.7}$ cannot be expressed as a terminating or repeating decimal and consequently is not rational. Therefore, there is only one choice that does not include one or the other or both of these two numbers. Option B is the correct answer.
9	D	$\sqrt{3}$ is non-terminating and non-repeating.

Question No.	Answer	Detailed Explanation
10	C	Step 1- Write equation 1 - Assign the repeating rational number to x: x = 0.1515... Step 2- Write equation 2 - Multiply equation 1 by 100: 100x = 15.1515... Step 3- Subtract equation 1 from 2: 100x = 15.1515... x = 0.1515... 99x = 15 x = 15/99
11	C	Step 1- Write equation 1 - Assign the repeating rational number to x: x = 0.112112... Step 2- Write equation 2 - Multiply equation 1 by 1000: 1000x = 112.112... Step 3- Subtract equation 1 from 2: 1000x = 112.112112... x = 0.112112... 999x = 112 x = 112/999
12	B	$\sqrt{2}$ cannot be expressed as a non-terminating or non-repeating decimal and is therefore irrational. However, it is real.
13	D	The square root of a negative number is not a real number. Since the first three choices contain real numbers, none of them will fit.
14	B	Numbers with terminating decimals are real (R) and rational (Q), but are not integers (Z), whole numbers (W), or natural numbers (N).
15	C	By definition all rational numbers may be written as terminating or repeating decimals. Irrational numbers can be written as decimals (non-repeating, non-terminating), but not as fractions.
16	A, C	Rational numbers are numbers that can be expressed as a fraction. Since 5/7 is already a fraction representing a rational number. The second answer, the square root of 25 gives you a whole number 5. This can be written as a fraction, 5/1. Thus, both are rational numbers.

Question No.	Answer	Detailed Explanation
17	<table><tr><td></td><td>Rational</td><td>Irrational</td></tr><tr><td>$\sqrt{2}$</td><td></td><td>●</td></tr><tr><td>$\frac{1}{3}$</td><td>●</td><td></td></tr><tr><td>0.575</td><td>●</td><td></td></tr><tr><td>$\frac{\sqrt{12}}{4}$</td><td></td><td>●</td></tr></table>	Rational numbers are numbers that can be expressed as a fraction. The square root of 2 gives you a decimal that does not repeat and doesn't end. Thus making it irrational. 1/3 is already expressed as a fraction, so it is rational. 0.575 can be written as 575/1000, thus making it rational. The last choice is a fraction, however, since it is a radical (and not a perfect square), this will not simplify into a fraction with integers as your numerator and denominator.
18	C	$\sqrt{10}$ is an irrational number because it cannot be written as a fraction. The others can.

Lesson 2: Approximating Irrational Numbers

Question No.	Answer	Detailed Explanation
1	B	$2^2=4$ and $3^2=9$ Since 5 lies between 4 and 9, $\sqrt{5}$ lies between 2 & 3.
2	C	$(2.2)^2=4.84$ and $(2.3)^2=5.29$ Since 5 lies between 4.84 and 5.29 on the number line, $\sqrt{5}$ lies between 2.2 and 2.3.
3	C	If we change the numbers all to the same accuracy, it is easier to order them on the number line. Write 1.8 as 1.80, 2.5 as 2.50 and $\sqrt{5}$ as >2 because $2^2=4$ and < 2.5 because $2.5^2=6.25$. Then, the correct order is 1.35, 1.8, $\sqrt{5}$, 2.5.
4	C	$2.63^2 \approx 6.92$ $2.65^2 \approx 7.02$ The square root of 7 is about 2.64, so $\sqrt{7}$ falls between 2.63 and 2.65. Then, 2.5, 2.63, $\sqrt{7}$, 2.65 is the correct answer.
5	B	$\sqrt{(0.6561)}=.81$ .8... means that the 8 is repeating; i.e. .888... .8 may be written as .80 So .8 represents the smallest (least) value.
6	A	$4\frac{1}{2} = 4.5$ $2<\sqrt{5}<3$ $\sqrt{10}>3$; so $2\sqrt{10}>6$ Then, the correct order is: $\sqrt{5}$, $4\frac{1}{2}$, 4.75, $2\sqrt{10}$
7	C	$\sqrt{.4}$ =(approx.) .63 which is the greatest of these numbers.
8	B	$\sqrt{.9}$ = (approx.) .95 So .9, $\sqrt{.9}$, .999, .9... is the correct order.
9	A	$\frac{32}{8} = 4$, $3\frac{7}{8} = 3.875$, π = (approx) 3.14, $\sqrt{2}$ = (approx) 1.4 The correct order is $\sqrt{2}$, π, $3\frac{7}{8}$, $\frac{32}{8}$
10	D	Convert each of the numbers to a similar expression. $\sqrt{12}$= approx. 3.46 $3\frac{7}{12}$= approx. 3.58 $3\frac{9}{10}$= 3.9 The correct order is $\sqrt{12}$, $3\frac{7}{12}$, 3.6, $3\frac{9}{10}$

Question No.	Answer	Detailed Explanation
11	B	$\sqrt{24}$ = approx. 4.90 $4\frac{7}{8}$ = 4.875 So $4\frac{7}{8}$ and 5.0 is the correct answer.
12	B	$1^2 = 1$ and $2^2 = 4$ Since 2 >1 and <4, $\sqrt{2}$ lies between 1 and 2.
13	B	$1.41^2 = 1.988$ and $1.42^2 = 2.0164$ Since 2 lies between 1.988 and 2.0164 on the number line, $\sqrt{2}$ lies between 1.41 and 1.42 on the number line.
14	A	$1^2 = 1$ and $2^2 = 4$ Since 3 lies between 1^2 and 2^2, Then $\sqrt{3}$ lies between 1 and 2.
15	D	$1.73^2 = 2.9929$ and $1.74^2 = 3.0276$ Since 3 lies between 2.9929 and 3.0276 on the number line, $\sqrt{3}$ lies between 1.73 and 1.74.

Chapter 2: Expressions and Equations

Lesson 1: Properties of Exponents

1. Is -5^2 equal to $(-5)^2$?

Ⓐ Yes, because they both equal -25.
Ⓑ Yes, because they both equal -10.
Ⓒ Yes, because they both equal 25.
Ⓓ No, because -5^2 equals -25 and $(-5)^2$ equals 25.

2. $\frac{X^6}{X^{-2}}$ =

Ⓐ $\frac{1}{X^3}$

Ⓑ $\frac{1}{X^{12}}$

Ⓒ X^4

Ⓓ X^8

3. Which of the following is equal to 3^{-2} ?

Ⓐ $\frac{1}{9}$

Ⓑ -9

Ⓒ 9

Ⓓ $\frac{1}{6}$

4. Which of the following is equivalent to $X^{(2-5)}$?

Ⓐ X^3

Ⓑ $X^{\frac{1}{3}}$

Ⓒ $\frac{1}{X^3}$

Ⓓ 3^X

5. $1^9 =$

Ⓐ 1
Ⓑ 3
Ⓒ 9
Ⓓ $\frac{1}{9}$

6. $(X^{-3})(X^{-3}) =$

Ⓐ X^6
Ⓑ X^9
Ⓒ $\frac{1}{X^6}$
Ⓓ $\frac{1}{X^9}$

7. $(X^{-2})^{-7} =$

Ⓐ X^5

Ⓑ X^{14}
Ⓒ $\frac{1}{X^5}$

Ⓓ $\frac{1}{X^{14}}$

8. $(X^4)^0 =$

Ⓐ X
Ⓑ X^4
Ⓒ 1
Ⓓ 0

9. $(3^2)^3 =$

Ⓐ 3^5
Ⓑ 3^6
Ⓒ 3
Ⓓ 1

10. $5^2+5^3 =$ _____

11. Which of the following show the proper laws of exponents?

Note : More than one option may be correct. Select all the correct answers.

Ⓐ $3^2 x 3^5=3^{10}$
Ⓑ $(4^2)^3 = 4^6$
Ⓒ $\frac{8^5}{8^1} = 8^4$
Ⓓ $7^4 x 7^4=7^8$

12. Simplify this expression.
$a^7(a^8)(a)$

Write your answer in the box below.

13. Select the ones that properly applied the different Laws of Exponents, making sure to keep positive exponents.

Note : More than one option may be correct. Select all the correct answers.

Ⓐ $(4a^3)^2 = 16a^6$
Ⓑ $(2x^4)^2 = 4x^6$
Ⓒ $(x^2y^{-1})^2 = \frac{x^4}{y^2}$
Ⓓ $(2a^{-2})^3 = 8a^6$

CHAPTER 2 → Lesson 2: Square & Cube Roots

1. What is the cube root of 1,000 ?

Ⓐ 10

Ⓑ 100

Ⓒ $33\frac{1}{3}$

Ⓓ $333\frac{1}{3}$

2. $8\sqrt{12} \div \sqrt{15} =$

Ⓐ $\frac{4}{5}$

Ⓑ $\frac{8}{5}$

Ⓒ $\frac{16}{\sqrt{5}}$

Ⓓ $\frac{\sqrt{5}}{8}$

3. The square root of 75 is between which two integers?

Ⓐ 8 and 9
Ⓑ 7 and 8
Ⓒ 9 and 10
Ⓓ 6 and 7

4. The square root of 110 is between which two integers?

Ⓐ 10 and 11
Ⓑ 9 and 10
Ⓒ 11 and 12
Ⓓ 8 and 9

5. Solve the following problem: $6\sqrt{20} \div \sqrt{5} =$ _____

Ⓐ 12
Ⓑ 11
Ⓒ 30
Ⓓ 5

6. **The cube root of 66 is between which two integers?**

Ⓐ 4 and 5
Ⓑ 3 and 4
Ⓒ 5 and 6
Ⓓ 6 and 7

7. **Which expression has the same value as $3\sqrt{144} \div \sqrt{12}$?**

Ⓐ $3\sqrt{12}$
Ⓑ $4\sqrt{12}$
Ⓒ $27 \div \sqrt{12}$
Ⓓ $33 \div \sqrt{12}$

8. **The cubic root of 400 lies between which two numbers?**

Ⓐ 5 and 6
Ⓑ 6 and 7
Ⓒ 7 and 8
Ⓓ 8 and 9

9. **Which of the following is equivalent to the expression $4\sqrt{250} \div 5\sqrt{2}$?**

Ⓐ $4\sqrt{25} \div 5$
Ⓑ $4\sqrt{125} \div \sqrt{2}$
Ⓒ $4\sqrt{10}$
Ⓓ $4\sqrt{5}$

10. **The cube root of 150 is closest to which of the following?**

Ⓐ 15
Ⓑ 10
Ⓒ 5
Ⓓ 3

11. **Select all that apply: What is the value of $\sqrt{\frac{81}{289}}$?**

Ⓐ $\frac{1}{2}$

Ⓑ $\frac{9}{17}$

Ⓒ $\frac{-1}{2}$

Ⓓ $\frac{-9}{17}$

12. **Select all the numbers which have integers as the cube roots.**

Ⓐ $\sqrt[3]{27}$

Ⓑ $\sqrt[3]{9}$

Ⓒ $\sqrt[3]{1000}$

Ⓓ $\sqrt[3]{18}$

13. **Fill in the boxes to make the statement true**

$\sqrt[3]{8} = \square$ since $\square \times \square \times \square = 8$

CHAPTER 2 → Lesson 3: Scientific Notation

1. **In 2007, approximately 3,380,000 people visited the Statue of Liberty. Express this number in scientific notation.**

   Ⓐ 0.388×10^7
   Ⓑ 3.38×10^6
   Ⓒ 33.8×10^5
   Ⓓ 338×10^4

2. **The average distance from Saturn to the Sun is 890,800,000 miles. Express this number in scientific notation.**

   Ⓐ 8908×10^8
   Ⓑ 8908×10^5
   Ⓒ 8.908×10^8
   Ⓓ 8.908×10^5

3. **The approximate population of Los Angeles is 3.8×10^6 people. Express this number in standard notation.**

   Ⓐ 380,000
   Ⓑ 3,800,000
   Ⓒ 38,000,000
   Ⓓ 380,000,000

4. **The approximate population of Kazakhstan is 1.53×10^7 people. Express this number in standard notation.**

   Ⓐ 153,000
   Ⓑ 1,530,000
   Ⓒ 15,300,000
   Ⓓ 153,000,000

5. **The typical human body contains about 2.5×10^{-3} kilograms of zinc. Express this amount in standard form.**

   Ⓐ 0.00025 kilograms
   Ⓑ 0.0025 kilograms
   Ⓒ 0.025 kilograms
   Ⓓ 0.25 kilograms

6. If a number expressed in scientific notation is N x 10^5, how large is the number?

Ⓐ Between 1,000 (included) and 10,000
Ⓑ Between 10,000 (included) and 100,000
Ⓒ Between 100,000 (included) and 1,000,000
Ⓓ Between 1,000,000 (included) and 10,000,000

7. Red light has a wavelength of 650 x 10^{-9} meters. Express the wavelength in scientific notation.

Ⓐ 65.0 x 10^{-10} meters
Ⓑ 65.0 x 10^{-8} meters
Ⓒ 6.50 x 10^{-7} meters
Ⓓ 6.50 x 10^{-11} meters

8. A strand of hair from a human head is approximately 1 x 10^{-4} meters thick. What fraction of a meter is this?

Ⓐ $\frac{1}{100}$
Ⓑ $\frac{1}{1,000}$
Ⓒ $\frac{1}{10,000}$
Ⓓ $\frac{1}{100,000}$

9. Which of the following numbers has the greatest value?

Ⓐ 8.93 x 10^3
Ⓑ 8.935 x 10^2
Ⓒ 8.935 x 10^3
Ⓓ 893.5 x 10^1

10. Which of the following numbers has the least value?

Ⓐ -1.56 x 10^2
Ⓑ -1.56 x 10^3
Ⓒ 1.56 x 10^2
Ⓓ 1.56 x 10^3

11. **Which of the following are correctly written in scientific notation?**

Note that more than one option may be correct. Select all the correct options

Ⓐ **$.032 \times 10^5$**
Ⓑ **11.002×10^{-1}**
Ⓒ **1.23×10^5**
Ⓓ **9.625×10^{-7}**

12. **Convert 2,347,000,000 from standard form to scientific notation by filling in the blank boxes.**

$\square \times 10^{\square}$

13. **Convert 0.0000687 to scientific notation by filling in the blank boxes.**

$\square \times 10^{\square}$

CHAPTER 2 → Lesson 4: Solving Problems Involving Scientific Notation

1. The population of California is approximately 3.7×10^7 people. The land area of California is approximately 1.6×10^5 square miles. Divide the population by the area to find the best estimate of the number of people per square mile in California.

   Ⓐ 24 people
   Ⓑ 240 people
   Ⓒ 2,400 people
   Ⓓ 24,000 people

2. Mercury is approximately 6×10^7 kilometers from the Sun. The speed of light is approximately 3×10^5 kilometers per second. Divide the distance by the speed of light to determine the approximate number of seconds it takes light to travel from the Sun to Mercury.

   Ⓐ 2 seconds
   Ⓑ 20 seconds
   Ⓒ 200 seconds
   Ⓓ 2,000 seconds

3. Simplify $(4 \times 10^6) \times (2 \times 10^3)$ and express the result in scientific notation.

   Ⓐ 8×10^9
   Ⓑ 8×10^{18}
   Ⓒ 6×10^9
   Ⓓ 6×10^{18}

4. Simplify $(2 \times 10^{-3}) \times (3 \times 10^5)$ and express the result in scientific notation.

   Ⓐ 5×10^{-8}
   Ⓑ 5×10^{-15}
   Ⓒ 6×10^8
   Ⓓ 6×10^2

5. Washington is approximately 2.4×10^3 miles from Utah. Mary drives 6×10^1 miles per hour from Washington to Utah. Divide the distance by the speed to determine the approximate number of hours it takes Mary to travel from Washington to Utah.

   Ⓐ 41 hours
   Ⓑ 40 hours
   Ⓒ 39 hours
   Ⓓ 38 hours

6. **Which of the following is NOT equal to $(5 \times 10^5) \times (9 \times 10^{-3})$?**

Ⓐ 4.5×10^4
Ⓑ 4.5×10^3
Ⓒ 4,500
Ⓓ 45×100

7. **Find $(5 \times 10^7) \div (10 \times 10^2)$ and express the result in scientific notation.**

Ⓐ 5×10^4
Ⓑ 0.5×10^5
Ⓒ 50×10^9
Ⓓ 5.0×10^9

8. **Approximate 0.00004567 x 0.00001234 and express the result in scientific notation.**

Ⓐ 5.636×10^{-8}
Ⓑ 5.636×10^{-9}
Ⓒ 5.636×10^{-10}
Ⓓ None of the above.

9. **Find the product $(50.67 \times 10^4) \times (12.9 \times 10^3)$ and express the answer in standard notation.**

Ⓐ 653.643
Ⓑ 65,364,300,000
Ⓒ 6.53643×10^9
Ⓓ 6,536,430,000

10 **Approximate the quotient and express the answer in standard notation.**

$(1.298 \times 10^4) \div (3.97 \times 10^2)$

Ⓐ 32.7
Ⓑ 0.327
Ⓒ 0.327×10^2
Ⓓ None of the above.

11. Which of the following examples correctly applies the rules of scientific notation? Select all the correct answers.

Ⓐ $(4.0 \times 10^3)(5.0 \times 10^5) = 2 \times 10^9$

Ⓑ $\dfrac{4.5 \times 10^5}{9.0 \times 10^9} = 2 \times 10^4$

Ⓒ $(2.1 \times 10^5) + (2.7 \times 10^5) = 4.8 \times 10^5$

Ⓓ $(3.1 \times 10^5) - (2.7 \times 10^2) = 0.4 \times 10^3$

12. Which of the following lists are ordered from greatest to least? Select all the correct answers.

Ⓐ 2.0×10^2, 3.0×10^6, 4.0×10^{-7}, 5.0×10^{12}

Ⓑ 4.0×10^{-7}, 2.0×10^2, 3.0×10^6, 5.0×10^{12}

Ⓒ 3.0×10^7, 3.0×10^6, 3.0×10^2, 3.0×10^{-7}

Ⓓ 1.8×10^9, 1.5×10^6, 1.4×10^{-5}, 1.9×10^{-8}

13. $(6 \times 10^3)(9.91 \times 10^0) =$ __________

CHAPTER 2 → Lesson 5: Compare Proportions

1. **Find the unit rate if 12 tablets cost $1,440.**

   Ⓐ $100
   Ⓑ $150
   Ⓒ $120
   Ⓓ $50

2. **A package of Big Bubbles Gum has 10 pieces and sells for $2.90. A package of Fruity Gum has 20 pieces and sells for $6.20. Compare the unit prices.**

   Ⓐ Big Bubbles is $0.10 more per piece than Fruity.
   Ⓑ Fruity is $0.02 more per piece than Big Bubbles.
   Ⓒ They both have the same unit price.
   Ⓓ It cannot be determined.

3. **The first major ski slope in Vermont has a rise of 9 feet vertically for every 54 feet horizontally. A second ski slope has a rise of 12 feet vertically for every 84 feet horizontally. Which of the following statements is true?**

   Ⓐ The first slope is steeper than the second.
   Ⓑ The second slope is steeper than the first.
   Ⓒ Both slopes have the same steepness.
   Ⓓ Cannot be determined from the information given.

4. **Which of the following ramps has the steepest slope?**

   Ⓐ Ramp A has a vertical rise of 3 feet and a horizontal run of 15 feet
   Ⓑ Ramp B has a vertical rise of 4 feet and a horizontal run of 12 feet
   Ⓒ Ramp C has a vertical rise of 2 feet and a horizontal run of 10 feet
   Ⓓ Ramp D has a vertical rise of 5 feet and a horizontal run of 20 feet

5. **Choose the statement that is true about unit rate.**

   Ⓐ The unit rate can also be called the rate of change.
   Ⓑ The unit rate can also be called the mode.
   Ⓒ The unit rate can also be called the frequency.
   Ⓓ The unit rate can also be called the median.

6. Which statement is false?

Ⓐ Unit cost is calculated by dividing the amount of items by the total cost.
Ⓑ Unit cost is calculated by dividing the total cost by the amount of items.
Ⓒ Unit cost is the cost of one unit item.
Ⓓ On similar items, a higher unit cost is not the better price.

7. Selena is preparing for her eighth grade graduation party. She must keep within the budget set by her parents. Which is the best price for her to purchase ice cream?

Ⓐ $3.99/ 24 oz carton
Ⓑ $4.80/ one-quart carton
Ⓒ $11.00 / one gallon tub
Ⓓ $49.60/ five gallon tub

8. Ben is building a ramp for his skate boarding club. Which of the following provides the least steep ramp?

Ⓐ 2 feet vertical for every 10 feet horizontal
Ⓑ 3 feet vertical for every 9 feet horizontal
Ⓒ 4 feet vertical for every 16 feet horizontal
Ⓓ 5 feet vertical for every 30 feet horizontal

9. Riley is shopping for tee shirts. Which is the most expensive (based on unit price per shirt)?

Ⓐ 5 tee shirts for $50.00
Ⓑ 6 tee shirts for $90.00
Ⓒ 2 tee shirts for $22.00
Ⓓ 4 tee shirts for $48.00

10. The swim team is preparing for a meet. Which of the following is Lindy's fastest time?

Ⓐ five laps in fifteen minutes
Ⓑ four laps in sixteen minutes
Ⓒ two laps in ten minutes
Ⓓ three laps in eighteen minutes

11. David is having a Super Bowl party and he needs bottled sodas. Which of the following purchases will give him the lowest unit cost?

Ⓐ $2.00 for a 6 pack
Ⓑ $6.00 for a 24 pack
Ⓒ $3.60 for a 12 pack
Ⓓ $10.00 for a 36 pack

12. A package of plain wafers has 20 per pack and sells for $2.40. A package of sugar-free wafers has 30 pieces and sells for $6.30. Compare the unit prices.

Ⓐ Each plain wafer is $0.17 more than a sugar-free wafer.
Ⓑ Each sugar-free wafer is $0.09 more than a plain wafer.
Ⓒ They both have the same unit price per wafer.
Ⓓ The relationship cannot be determined.

13. Li took 4 practice tests to prepare for his chapter test.
Which of the following is the best score?

Ⓐ 36 correct out of 40 questions
Ⓑ 24 correct out of 30 questions
Ⓒ 17 correct out of 25 questions
Ⓓ 15 correct out of 20 questions

14. Mel's class is planning a fundraiser. They have decided to have a carnival. If they sell tickets in packs of 40 for $30.00, what is the unit cost per ticket?

Ⓐ $4.00 per ticket
Ⓑ $0.50 per ticket
Ⓒ $1.75 per ticket
Ⓓ $0.75 per ticket

15. Which of the following ski slopes has the steepest slope?

Ⓐ Ski Slope A has a vertical rise of 4 feet and a horizontal run of 16 feet
Ⓑ Ski Slope B has a vertical rise of 3 feet and a horizontal run of 12 feet
Ⓒ Ski Slope C has a vertical rise of 3 feet and a horizontal run of 9 feet
Ⓓ Ski Slope D has a vertical rise of 5 feet and a horizontal run of 25 feet

16. Solve for the proportion for the missing number.

$$\frac{2}{7} = \frac{4}{\square}$$

Fill in the blank box with the correct answer.

17. Solve the proportion to find the missing number.

$$\frac{20}{\square} = \frac{16}{20}$$

Fill in the blank box with the correct answer.

18. Write a proportion to solve the word problem below, then solve it.

Kasey bought 32 kiwi fruit for $16. How many kiwi can Lisa buy if she has $4?

CHAPTER 2 → Lesson 6: Understanding Slope

1. **Which of the following statements is true about slope?**

   Ⓐ Slopes of straight lines will always be positive numbers.
   Ⓑ The slopes vary between the points on a straight line.
   Ⓒ Slope is determined by dividing the horizontal distance between two points by the corresponding vertical distance.
   Ⓓ Slope is determined by dividing the vertical distance between two points by the corresponding horizontal distance.

2. **Which of the following is an equation of the line passing through the points (-1, 4) and (1, 2)?**

   Ⓐ $y = x - 3$
   Ⓑ $y = 2x + 2$
   Ⓒ $y = -2x + 4$
   Ⓓ $y = -x + 3$

3. **The graph of which equation has the same slope as the graph of $y = 4x + 3$?**

   Ⓐ $y = -2x + 3$
   Ⓑ $y = 2x - 3$
   Ⓒ $y = -4x + 2$
   Ⓓ $y = 4x - 2$

4. **Which of these lines has the greatest slope?**

   Ⓐ $y = \frac{8}{5}x - 7$
   Ⓑ $y = \frac{6}{5}x + 4$
   Ⓒ $y = \frac{7}{5}x + 2$
   Ⓓ $y = \frac{9}{5}x - 3$

5. **Which of these lines has the smallest slope?**

   Ⓐ $y = \frac{1}{8}x + 7$
   Ⓑ $y = \frac{1}{3}x + 7$
   Ⓒ $y = \frac{1}{4}x - 9$
   Ⓓ $y = \frac{1}{7}x$

6. **Fill in the blank with one of the four choices to make the following a true statement. Knowing _____ and the y-intercept is NOT enough for us to write the equation of the line.**

Ⓐ direction
Ⓑ a point on a given line
Ⓒ the x-intercept
Ⓓ the slope

7. **A skateboarder is practicing at the city park. He is skating up and down the steepest straight line ramp. If the highest point on the ramp is 30 feet above the ground and the horizontal distance from the base of the ramp to a point directly beneath the upper end is 500 feet, what is the slope of the ramp?**

Ⓐ $\frac{500}{30}$
Ⓑ $\frac{50}{3}$
Ⓒ $\frac{3}{50}$
Ⓓ None of these.

8. **If the equation of a line is expressed as $y = \frac{3}{2}x - 9$, what is the slope of the line?**

Ⓐ - 9

Ⓑ +9

Ⓒ $\frac{3}{2}$

Ⓓ $\frac{2}{3}$

9. **Which of the following is an equation of the line that passes through the points (0, 5) and (2, 15)?**

Ⓐ y = 5x + 5
Ⓑ y = 5x + 3
Ⓒ y = 3x + 5
Ⓓ y = 5x - 5

10. Which of the following equations has the same slope as the line passing through the points (1, 6) and (3, 10)?

Ⓐ y = 2x - 9
Ⓑ y = 5x - 2
Ⓒ y = 4x - 5
Ⓓ y = 9x - 6

11. Which of the following equations has the same slope as the line passing through the points (3, 6) and (5, 10)?

Ⓐ y = 2x -12
Ⓑ y = 11x - 8
Ⓒ y = -2x - 9
Ⓓ y = 3x - 5

12. Which equation has the same slope as y = -5x - 4?

Ⓐ y = 5x +15
Ⓑ y = -5x - 11
Ⓒ y = 5x -19
Ⓓ y = 5x - 13

13. Find the slope of the line passing through the points (3,3) and (5,5).

Ⓐ 2
Ⓑ 1
Ⓒ 3
Ⓓ 5

14. Which of the following lines has the steepest slope?

Ⓐ y = 4x+5
Ⓑ y =-3x + 5
Ⓒ y = 3x - 5
Ⓓ They all have the same slope.

15. Which of the following is the equation of the line passing through the points (0,-3) and (-3,0) ?

Ⓐ y = -3x + 3
Ⓑ y = -3x
Ⓒ y = -x - 3
Ⓓ y = -x

16. Find the slope between the points (-12, -5) and (0, 8).

Write your answer in the box given below.

17. Find the slope between the points (3, -3) and (12, -2).

Write your answer in the box given below.

18. Find the slope between the points (1, 2) and (5, -7).

Write your answer in the box given below.

CHAPTER 2 → Lesson 7: Solving Linear Equations

1. Which two consecutive odd integers have a sum of 44?

Ⓐ 21 and 23
Ⓑ 19 and 21
Ⓒ 23 and 25
Ⓓ 17 and 19

2. During each of the first three quarters of the school year, Melissa earned a grade point average of 2.1, 2.9, and 3.1. What does her 4th quarter grade point average need to be in order to raise her grade to a 3.0 cumulative grade point average?

Ⓐ 3.9
Ⓑ 4.2
Ⓒ 2.6
Ⓓ 3.5

3. Martha is on a trip of 1,924 miles. She has already traveled 490 miles. She has 3 days left on her trip. How many miles does she need to travel each day to complete her trip?

Ⓐ 450 miles/day
Ⓑ 464 miles/day
Ⓒ 478 miles/day
Ⓓ 492 miles/day

4. Find the solution to the following equation: $3x + 5 = 29$

Ⓐ $x = 24$
Ⓑ $x = 11$
Ⓒ $x = 8$
Ⓓ $x = 6$

5. Find the solution to the following equation:
$7 - 2x = 13 - 2x$

Ⓐ $x = -10$
Ⓑ $x = -3$
Ⓒ $x = 3$
Ⓓ There is no solution.

6. Find the solution to the following equation: 6x + 1 = 4x - 3

Ⓐ x = -1
Ⓑ x = -2
Ⓒ x = - 0.5
Ⓓ There is no solution.

7. Find the solution to the following equation:
2x + 6 + 1 = 7 + 2x

Ⓐ x = -3
Ⓑ x = 3
Ⓒ x = 7
Ⓓ All real numbers are solutions.

8. Find the solution to the following equation: 8x-1=8x

Ⓐ x = 7
Ⓑ x = -8
Ⓒ x = 8
Ⓓ There is no solution.

9. Which of the answers is the correct solution to the following equation?
2x + 5x - 9 = 8x - x - 3 - 6

Ⓐ x = 3
Ⓑ x = 7
Ⓒ x = 9
Ⓓ All real values for x are correct solutions.

10. Solve the following linear equation for y.
-y + 7y -54 = 0

Ⓐ y = 0
Ⓑ y = 1
Ⓒ y = 6
Ⓓ y = 9

11. Solve each equation for the variable. Select the equations whose variable values are the same.

Note that more than one option may be correct. Select all the correct answers.

Ⓐ -5m = 25
Ⓑ -10c = -80
Ⓒ -7 + g = -12
Ⓓ 12m + 20 = -40

12. Solve for x: 5x + 20 = -20.

x = ?

Write your answer in the box given below.

13. Fill in the missing number to make this equation true, if x = -3.

7+3x=5x+______

CHAPTER 2 → Lesson 8: Solve Linear Equations with Rational Numbers

1. Solve the following linear equation: $\frac{7}{14} = n + \frac{7}{14}n$

Ⓐ $n = 1\frac{1}{2}$

Ⓑ $n = 3$

Ⓒ $n = \frac{1}{3}$

Ⓓ $n = 1$

2. Find the solution to the following equation: $2(2x - 7) = 14$

Ⓐ $x = 14$
Ⓑ $x = 7$
Ⓒ $x = 1$
Ⓓ $x = 0$

3. Solve the following equation for x.
$6x - (2x + 5) = 11$

Ⓐ $x = -3$
Ⓑ $x = -4$
Ⓒ $x = 3$
Ⓓ $x = 4$

4. $4x + 2(x - 3) = 0$

Ⓐ $x = 0$
Ⓑ $x = 1$
Ⓒ $x = 2$
Ⓓ The equation is true for all real values of x

5. Solve the following equation for y.
$3y - 7(y + 5) = y - 35$

Ⓐ $y = 0$
Ⓑ $y = 1$
Ⓒ $y = 2$
Ⓓ The equation is true for all real values of y

6. Solve the following linear equation: $2(x-5) = \frac{1}{2}(6x+4)$

Ⓐ $x= -12$
Ⓑ $x= -9$
Ⓒ $x= -4$
Ⓓ There is no solution.

7. Solve the following linear equation for x.

$3x + 2 + x = \frac{1}{3}(12x + 6)$

Ⓐ $x= -4$
Ⓑ $x= 2$
Ⓒ There is no solution.
Ⓓ All real values for x are correct solutions.

8. $\frac{1}{2}x + \frac{2}{3}x + 5 = \frac{5}{2}x + 6$

Ⓐ $x = \frac{33}{4}$

Ⓑ $x = \frac{1}{2}$

Ⓒ $x = -\frac{3}{4}$

Ⓓ $x = -\frac{6}{5}$

9. Which of the following could be a correct procedure for solving the equation below?
$2(2x+3) = 3(2x+5)$

Ⓐ $4x+5 = 6x+5$
$-2x+5 = 5$
$-2x = 0$
$x = 0$

Ⓑ $4x+6 = 6x+15$
$-2x+6 = 15$
$-2x = 9$
$x = -\frac{9}{2}$

Ⓒ $2(5x) = 6x+15$
$10x = 6x+15$
$4x = 15$
$x = \frac{15}{4}$

Ⓓ $4x+6 = 6x+15$
$-2x+6 = 15$
$-2x = 9$
$x = \frac{2}{9}$

10. Solve the following linear equation:
$0.64x - 0.15x + 0.08 = 0.09x$

Ⓐ $x = -5$
Ⓑ $x = -0.2$
Ⓒ $x = 5.125$
Ⓓ There is no solution.

11. Which of the following statements are true? Select all that apply.

Ⓐ $w - \frac{2}{5} = \frac{8}{5}$ so $w = 2$

Ⓑ $\frac{-5}{8}y = 15$ so $y = 24$

Ⓒ $0.4x - 1.2 = 0.15x + 0.8$ so $x = 8$

Ⓓ $\frac{x}{6} = -5$ so $x = 30$

12. Select the answer choice with the correct order of how you would solve the equation.

$$\frac{3}{4}+\frac{1}{2}(m+\frac{1}{4})=\frac{19}{16}$$

X --> $m=\frac{5}{8}$ Y --> $\frac{1}{2}(m+\frac{1}{4})=\frac{7}{16}$ Z --> $(m+\frac{1}{4})=\frac{7}{8}$

Ⓐ X, Y, Z
Ⓑ Y, Z, X
Ⓒ X, Z, Y
Ⓓ Z, X, Y

13. Solve the following linear equation:

$$\frac{8}{16}= n+\frac{8}{16}n$$

Date of Completion:______________ Score:______________

CHAPTER 2 → Lesson 9: Solutions to Systems of Equations

1. Which of the following points is the intersection of the graphs of the lines given by the equations y = x - 5 and y = 2x + 1 ?

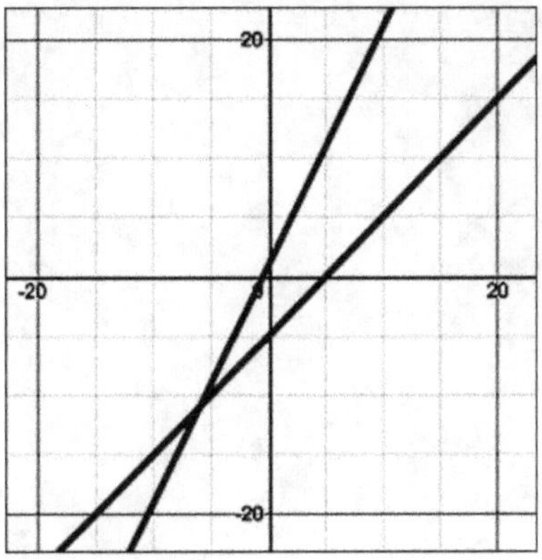

Ⓐ (1, 3)
Ⓑ (-1, -4)
Ⓒ (-2, -3)
Ⓓ (-6, -11)

2. Which of the following describes the solution set of this system?
y = 0.5x + 7
y = 0.5x - 1

Ⓐ The solution is (-2, -3) because the graphs of the two equations intersect at that point.
Ⓑ The solution is (0.5, 3) because the graphs of the two equations intersect at that point.
Ⓒ There is no solution because the graphs of the two equations are parallel lines.
Ⓓ There are infinitely many solutions because the graphs of the two equations are the same line.

3. Find the solution to the following system:
y = 2(2 - 3x)
y = -3(2x + 3)

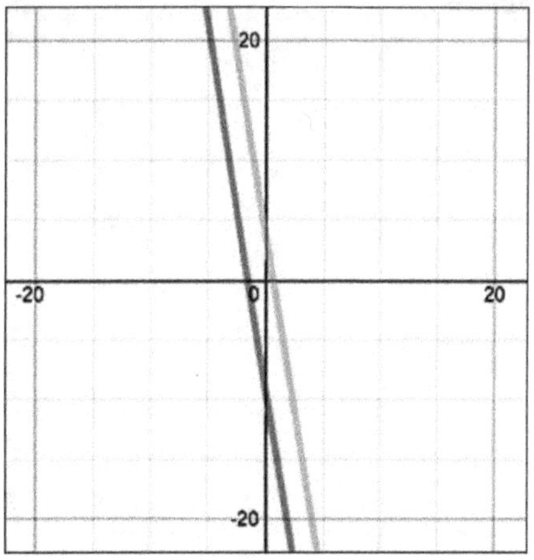

Ⓐ x = -1; y = 10
Ⓑ x = -2; y = 24
Ⓒ x = -3; y = 22
Ⓓ There is no solution.

4. **Use the graph to find the solution to the following system:**

$\frac{x}{2} + \frac{y}{3} = 2$

$3x - 2y = 48$

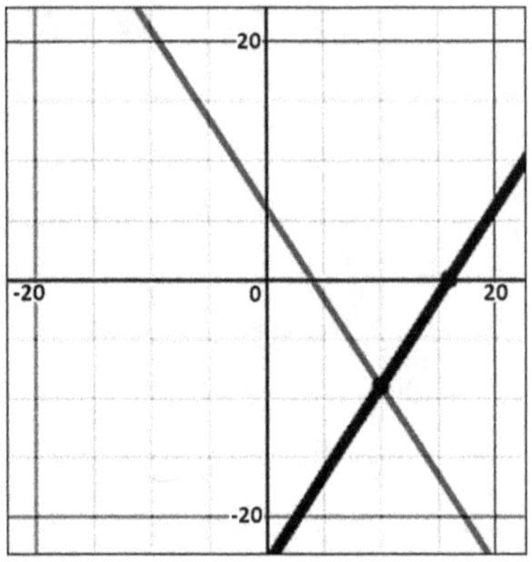

Ⓐ $x = 8, y = -6$
Ⓑ $x = 10, y = -9$
Ⓒ $x = 12, y = -3$
Ⓓ $x = 16, y = 0$

5. **Which of the following best describes the relationship between the graphs of the equations in this system?**
$y = 2x - 6$
$y = -2x + 6$

Ⓐ The lines intersect at the point (0, -3).
Ⓑ The lines intersect at the point (3, 0).
Ⓒ The lines do not intersect because their slopes are opposites and their y-intercepts are opposites.
Ⓓ They are the same line because their slopes are opposites and their y-intercepts are opposites.

6. **Solve the system:**

$2x + 3y = 14$
$2x - 3y = -10$

Ⓐ $x = 1, y = 4$
Ⓑ $x = 2, y = 12$
Ⓒ $x = 4, y = 2$
Ⓓ $x = 10, y = 10$

7. **Solve the system:**

$x = 13 + 2y$
$x - 2y = 13$

Ⓐ $x = 0, y = 13$
Ⓑ $x = 13, y = 0$
Ⓒ There is no solution.
Ⓓ There are infinitely many solutions.

8. **Solve the system:**

$y = 3x - 7$
$x + y = 9$

Ⓐ $x = 3, y = 6$
Ⓑ $x = 4, y = 5$
Ⓒ $x = 5, y = 4$
Ⓓ $x = 6, y = 3$

9. **Solve the system:**

$2x + 5y = 12$
$2x + 5y = 9$

Ⓐ $x = 1, y = 2$
Ⓑ $x = 2, y = 1$
Ⓒ There is no solution.
Ⓓ There are infinitely many solutions.

10. Solve the system:

$-4x + 7y = 26$
$4x + 7y = 2$

Ⓐ x = -3, y = 2
Ⓑ x = 3, y = -2
Ⓒ x = -2, y = 3
Ⓓ x = 2, y = -3

11. Randy has to raise $50.00 to repair his bicycle. He is only $1.00 short. He has only $1 and $5 bills. If he has one more $1 bills than $5 bills, how many does he have of each?

Ⓐ Ten $1-bills, Nine $5-bills
Ⓑ Nine $1-bills and Eight $5-bills
Ⓒ Eight $1-bills and Seven $5-bills
Ⓓ Seven $1-bills and Six $5-bills

12. Anya is three years older than her brother, Cole. In 11 years, Cole will be twice Anya's current age. Find their current ages.

Ⓐ Anya: 11 years old, Cole: 8 years old
Ⓑ Anya: 10 years old, Cole: 7 years old
Ⓒ Anya: 9 years old, Cole: 6 years old
Ⓓ Anya: 8 years old, Cole: 5 years old

CHAPTER 2 → Lesson 10: Solving Systems of Equations

1. **Find the solution to the following system of equations:**
   $13x + 3y = 15$ and $y = 5 - 4x$.

   Ⓐ $x = 0, y = 5$
   Ⓑ $x = 5, y = 0$
   Ⓒ $x = 9, y = -31$
   Ⓓ All real numbers are solutions.

2. **Solve the system:**
   $y = 2x + 5$
   $y = 3x - 7$

   Ⓐ $x = 12, y = 29$
   Ⓑ $x = 3, y = 11$
   Ⓒ $x = 5, y = -2$
   Ⓓ $x = -1, y = 3$

3. **Solve the system:**
   $2x + 3y = 14$
   $2x - 3y = -10$

   Ⓐ $x = 1, y = 4$
   Ⓑ $x = 2, y = 12$
   Ⓒ $x = 4, y = 2$
   Ⓓ $x = 10, y = 10$

4. **Solve the system:**
   $x = 13 + 2y$
   $x - 2y = 13$

   Ⓐ $x = 0, y = 13$
   Ⓑ $x = 13, y = 0$
   Ⓒ There is no solution.
   Ⓓ There are infinitely many solutions.

5. **Solve the system:**
   $2x + 5y = 12$
   $2x + 5y = 9$

   Ⓐ $x = 1, y = 2$
   Ⓑ $x = 2, y = 1$
   Ⓒ There is no solution.
   Ⓓ There are infinitely many solutions.

6. **Solve the system:**
-4x + 7y = 26
4x + 7y = 2

Ⓐ x = -3, y = 2
Ⓑ x = 3, y = -2
Ⓒ x = -2, y = 3
Ⓓ x = 2, y = -3

7. **Find the solution to the following system:**
y + 3x = 11
y - 2x = 1

Ⓐ x = -5, y = 2
Ⓑ x = 2, y = 5
Ⓒ x = -2, y = -5
Ⓓ x = -2, y = 5

8. **Solve the system:**
2x + 4y = 14
x + 2y = 7

Ⓐ x = -1, y = 4
Ⓑ x = 1, y = 3
Ⓒ There is no solution.
Ⓓ There are infinitely many solutions.

9. **Solve the system:**
3(y - 2x) = 9
x - 4 = 0

Ⓐ x = -4, y = -5
Ⓑ x = 4, y = 11
Ⓒ There is no solution.
Ⓓ There are infinitely many solutions.

10. Solve the system:
$10x = -5(y+2)$
$y = 3x-7$

Ⓐ $x = -1, y = 4$
Ⓑ $x = 2, y = 2$
Ⓒ $x = 2, y = -2$
Ⓓ $x = 1, y = -4$

11. Which of these will have one solution?

Note that more than one option may be correct. Select all the correct answers.

Ⓐ $y = \frac{3}{4}x+1$
$y = -\frac{1}{2}x-4$

Ⓑ $y=-3x+2$
$3x+y=-4$

Ⓒ $y = \frac{1}{3}x -3$
$2x+y=4$

Ⓓ $-x+2y=-2$
$4y=2x-4$

12. Fill in the table with correct solution for each system of equations.

SYSTEM	SOLUTION
$y = \frac{1}{2}x - 1$ $y = -\frac{1}{4}x - 4$	
$y = 2x + 4$ $y = -3x - 1$	
$y = 4$ $y = 7x - 3$	
$y = -\frac{2}{3}x - 4$ $y = \frac{5}{3}x + 3$	

13. Select the systems that have no solution.

Note that more than one option may be correct. Select all the correct answers.

Ⓐ $y= -4x+7$
$y=-3x+3$

Ⓑ $y=\frac{3}{4}x-3$
$y=\frac{3}{4}x+2$

Ⓒ $y= x-2$
$y=x+2$

Ⓓ $y=2x+3$
$4x-2y=8$

Ⓔ $y+2x=-12$
$y=x+15$

14. Fill in the table with the correct number of solutions for each system of equations.

SYSTEM	NUMBER OF SOLUTIONS
$\begin{cases} -x + 2y = 14 \\ x - 2y = -11 \end{cases}$	
$\begin{cases} 2x + 5y = 5 \\ -2x - y = -23 \end{cases}$	
$\begin{cases} y = 3x + 6 \\ -6x+2y = 12 \end{cases}$	

CHAPTER 2 → Lesson 11: Systems of Equations in Real-World Problems

1. Jorge and Jillian have cell phones with different service providers. Jorge pays \$50 a month and \$1 per text message sent. Jillian pays \$72 a month and \$0.12 per text message sent. How many texts would each of them have to send for their bill to be the same amount at the end of the month?

   Ⓐ 2 texts
   Ⓑ 22 texts
   Ⓒ 25 texts
   Ⓓ 47 texts

2. Mr. Stevens is 63 years older than his grandson, Tom. In 3 years, Mr. Stevens will be four times as old as Tom. How old is Tom?

   Ⓐ 17 years
   Ⓑ 18 years
   Ⓒ 20 years
   Ⓓ 22 years

3. Janet has packed a total of 50 textbooks and workbooks in a box, but she can't remember how many of each are in the box. Each textbook weighs 2 pounds, and each workbook weighs 0.5 pounds, and the total weight of the books in the box is 55 pounds. If t is the number of textbooks and w is the number of workbooks, which of the following systems of equations represents this situation?

   Ⓐ $t + w = 55$
   $2t + 0.5w = 50$

   Ⓑ $2t + w = 50$
   $t + 0.5w = 55$

   Ⓒ $t + w = 50$
   $2t + 0.5w = 55$

   Ⓓ $t + w = 55$
   $2.5(t + w) = 50$

4. Plumber A charges \$50 to come to your house, plus \$40 per hour of labor. Plumber B charges \$75 to come to your house, plus \$35 per hour of labor. If y is the total dollar amount charged for x hours of labor, which of the following systems of equations correctly represents this situation?

Ⓐ $y = 50x + 40$
$y = 75x + 35$

Ⓑ $y = 50x + 40$
$y = 35x + 75$

Ⓒ $y = 40x + 50$
$y = 75x + 35$

Ⓓ $y = 40x + 50$
$y = 35x + 75$

5. 10 tacos and 6 drinks cost \$19.50. 7 tacos and 5 drinks cost \$14.25. If t is the cost of one taco and d is the cost of one drink, which of the following systems of equations represents this situation?

Ⓐ $10t + 6d = 19.50$
$7t + 5d = 14.25$

Ⓑ $6t + 10d = 19.50$
$5t + 7d = 14.25$

Ⓒ $10t + 7t = 19.50$
$6d + 5d = 14.25$

Ⓓ $16(t + d) = 19.50$
$12(t + d) = 14.25$

6. Cindy has \$25 saved and earns \$12 per week for walking dogs. Mindy has \$55 saved and earns \$7 per week for watering plants. Cindy and Mindy save all of the money they earn and do not spend any of their savings. After how many weeks will they have the same amount saved? How much money will they have saved?

Ⓐ After 4 weeks, they each will have \$83 saved.
Ⓑ After 5 weeks, they each will have \$85 saved.
Ⓒ After 6 weeks, they each will have \$97 saved.
Ⓓ After 7 weeks, they each will have \$104 saved.

7. The seventh and eighth grade classes are raising money for a field trip. The seventh graders are selling calendars for $1.50 each and the eighth graders are selling candy bars for $1.25 each. If they have sold a combined total of 1100 items and each class has the same income, find the number of each item that has been sold.

Ⓐ 400 calendars and 700 candy bars
Ⓑ 700 calendars and 400 candy bars
Ⓒ 500 calendars and 600 candy bars
Ⓓ 600 calendars and 500 candy bars

8. Anya is three years older than her brother, Cole. In 11 years, Cole will be twice Anya's current age. Find their current ages.

Ⓐ Anya: 11 years old
Cole: 8 years old
Ⓑ Anya: 10 years old
Cole: 7 years old
Ⓒ Anya: 9 years old
Cole: 6 years old
Ⓓ Anya: 8 years old
Cole: 5 years old

9. Tom and his sister both decided to get part-time jobs after school at competing clothing stores. Tom makes $15 an hour and receives $3 in commission for every item he sells. His sister makes $7 an hour and receives $5 in commission for every item she sells. How many items would each of them have to sell to make the same amount of money in an hour?

Ⓐ 1 item
Ⓑ 2 items
Ⓒ 3 items
Ⓓ 4 items

10. Lucia and Jack are training for a marathon. Lucia started the first day by running 2 miles and adds 0.25 mile to her distance every day. Jack started the first day by running 0.5 mile and adds 0.5 mile to his distance every day. If both continue this plan, on what day will Lucia and Jack run the same distance?

Ⓐ Day 3
Ⓑ Day 7
Ⓒ Day 9
Ⓓ Day 12

11. The admission fee at a carnival is \$3.00 for children and \$5.00 for adults. On the first day 1,500 people enter the fair and \$5740 is collected. How many children and how many adults attended the carnival?

Select the correct system and answer. There can be more than one correct answer, choose all applicable ones.

Ⓐ $\begin{cases} 3c + 5a = 1500 \\ c + a = 5740 \end{cases}$

Ⓑ $\begin{cases} 3c + 5a = 5740 \\ c + a = 1500 \end{cases}$

Ⓒ a = 620, c = 880

Ⓓ a = 936, c = 564

12. Match each correct solution to its corresponding word problem.

	15 and 19	\$1.50 and \$1.05	\$1.75 and \$1.60	18 and 31
1. You buy 5 bags of chips and 9 bags of pretzels for \$16.95. Later you buy 10 bags of chips and 10 bags of pretzels for \$25.50. Find the cost of 1 bag of chips and 1 bag of pretzels.				
2. You empty your coin jar and find 49 coins (all nickels and quarter). The total value of the coins is \$8.65. Find the number of nickels and quarters.				
3. Jill bought one hot dog and two soft drinks for a cost of \$4.95. Jack bought three hot dogs and one soft drink for a cost of \$6.85. Find the cost of one hot dog and one soft drink.				
4. There are a total of 34 lions and hyenas. Each lion eats 4 antelope. Each hyena eats 3 antelope. 117 antelope are eaten. Find the number of lions and hyenas.				

End of Expressions and Equations

ANSWER KEY AND DETAILED EXPLANATION

Chapter 2
Expressions and Equations

Lesson 1: Properties of Exponents

Question No.	Answer	Detailed Explanation
1	D	Unless there are parentheses to denote otherwise, the exponent is only applied to the constant or variable immediately preceding it. If there are parentheses immediately preceding it, then it is applied to everything within the parentheses.
2	D	When dividing quantities with like bases, you must subtract the exponents. 6 - (-2) =8 $\frac{X^6}{X^{-2}} = X^{6-(-2)} = X^8$
3	A	A quick way to change exponents from negative to positive is to move the expression to which the negative exponent is applied from the denominator to the numerator or vice versa and change the sign of the exponent. 3^{-2} is the same as $\frac{1}{3^2}$ which is the same as $\frac{1}{9}$.
4	C	$X^{(2-5)} = X^{-3}$ Now move x^{-3} to the denominator and change the sign of the exponent from negative to positive. $\frac{1}{X^3}$
5	A	Regardless of the number of 1s that we multiply the result is always 1 because 1 is the identity element for multiplication.
6	C	When multiplying quantities with the same base, you add exponents. $(X^{-3})(X^{-3}) = X^{-6}$ To change the exponent -6 to positive 6, you write the reciprocal of X^{-6}. $\frac{1}{X^6}$
7	B	$(x^{-2})^{-7} = x^{14}$ because to raise a power to a power, we multiply exponents.

Question No.	Answer	Detailed Explanation
8	C	4 x 0 = 0 and any number (other than 0, as 0^0 is not defined) to the 0 power is 1 by definition.
9	B	When raising a power to a power, multiply exponents.
10	150	$5^2 = 25$ and $5^3 = 125$ 25 +125 = 150
11	B,C & D	Option B is using the Power of a Power Property $(4^2)^3=4^6$ because $4^2 \times 4^2\ 4^2 = 4^{2+2+2}= 4^6$. Option C is using Division Property of Exponents $8^5/8^1=8^4$. This is true because $\frac{8\times8\times8\times8\times8}{8}$ when simplified, leaves you with 8×8×8×8 which is 8^4. Option D is using the Multiplicative Property of Exponents. $7^4 \times 7^4 = 7^8$ because $7\times7\times7\times7\times7\times7\times7\times7=7^8$
12	a^{16}	Using the product rule, when the base is the same, add the exponents. Here, 7 + 8 + 1 = 16, so the expression becomes a^{16}.
13	A, C	In the expression $(4a^3)^2$ you are applying the Power of a Power Law. You distribute the squared to each term in the parentheses to end up with 4^2a^{3x2} which simplifies to $16a^6$. The same thing applies to $(x^2y^{-1})^2$ so you end up with x^{2x2} and y^{-1x2}. These simplify to x^4 and y^{-2}. In this case, since you want to keep positive exponents, you have to move the y term to the denominator to keep positive exponents. Thus making it $\frac{x^4}{y^2}$

Lesson 2: Square & Cube Roots

Question No.	Answer	Detailed Explanation
1	A	$10^3 = 1,000$, So, the cube root of 1,000 is 10.
2	C	$8\sqrt{12} = 8\sqrt{(4 \times 3)} = 8(2)\sqrt{3} = 16\sqrt{3}$ $\sqrt{15} = \sqrt{(5 \times 3)}$ So $8\sqrt{12} \div \sqrt{15} = 16\sqrt{3} \div \sqrt{5}\sqrt{3}$
3	A	$8^2 = 64$ and $9^2 = 81$ Therefore, the square root of 75 is between 8 and 9.
4	A	$10^2 = 100$ and $11^2 = 121$ Therefore, the square root of 110 is between 10 and 11.
5	A	$6\sqrt{20} \div \sqrt{5} = 6\sqrt{4}\sqrt{5} \div \sqrt{5} = 6\sqrt{4} = 12$
6	A	$4^3 = 64$ and $5^3 = 125$ Therefore, the cube root of 66 is between 4 and 5.
7	A	$3\sqrt{144} \div \sqrt{12} = 3(12) \div \sqrt{12} = (3 \times \sqrt{12} \times \sqrt{12}) \div \sqrt{12} =$ $3\sqrt{12}$
8	C	$7^3 = 343$ and $8^3=512$ Therefore, the cube root of 400 lies between 7 and 8.
9	D	$4\sqrt{250} \div 5\sqrt{2} = 4\sqrt{25}\sqrt{10} \div 5\sqrt{2} = 4(5)\sqrt{10} \div 5\sqrt{2} = 4\sqrt{10}$ $\div \sqrt{2} = 4\sqrt{5}\sqrt{2} \div \sqrt{2} = 4\sqrt{5}$
10	C	$15^3 = 3,375$; $10^3 = 1,000$; $5^3 = 125$; $3^3 = 27$ 150 is closest to 5^3 or 125.
11	B, D	Both the options (B) and (D) are correct because if you take the square root of 81 it can be either -9 or 9. The same applies to fractions. When you take the square root of both the numerator and the denominator you have to consider that the factors can be either positive or negative.
12	A, C	Options (A) and (C) are correct, because if you take the cube root of them, you will get an integer. 27 can be written as 3 x 3 x 3 = 3^3 and 1000 can be written as 10 x 10 x 10=10^3. $\sqrt[3]{27} = 3$ $\sqrt[3]{1000} = 10$ 27, 1000 are called perfect cubes.
13	$\sqrt[3]{8}=2$ since $2 \times 2 \times 2 = 8$	Taking the cube root of a number is finding a number that multiplied by itself 3 times will give you the number under the radical. In this case the cube root of 8 will equal 2 because 2 x 2 x 2 = 8.

Lesson 3: Scientific Notation

Question No.	Answer	Detailed Explanation
1	B	Moving the decimal in 3,380,000 just to the right of the first non-zero digit requires us to move it 6 places to the left resulting in 3.38×10^{6}.
2	C	Moving the decimal in 890,800,000 just to the right of the first non-zero digit requires us to move it 8 places to the left resulting in 8.908×10^{8}.
3	B	Moving the decimal to the right 6 places from 3.8×10^{6}, we get 3,800,000.
4	C	From 1.53×10^{7}, we move the decimal to the right 7 places giving us 15,300,000.
5	B	Move the decimal 3 places to the left.
6	C	$1 \leq N \leq 9.99...$ and we must move the decimal 5 places to the right resulting in $100,000 \leq N \times 10^{5} < 1,000,000$.
7	C	To change to standard notation from 650×10^{-9}, we must move the decimal 9 places to the left because the exponent is negative. The result is 0.000000650. Changing to scientific notation, we get 6.5×10^{-7}. Or, since we must move the decimal 2 places to the left to get it properly placed for scientific notation, we only need to move it 7 more places for standard notation; so scientific notation would be written as 6.50×10^{-7}. Alternate method : $650 \times 10^{-9} = 6.50 \times 10^{2} \times 10^{-9}$. Using laws of exponents, we get $10^{2} \times 10^{-9} = 10^{-7}$. Therefore, wavelength of red light in scientific notation is 6.50×10^{-7} m
8	C	$1 \times 10^{4} = .0001$ (By moving decimal 4 places to the left because the exponent is -4. Since the 1 is in the ten-thousandths place, the fraction will be $\frac{1}{10,000}$
9	C, D	$8.935 \times 10^{3} = 8,935$ which is the greatest of the numbers shown.
10	B	Since we have both negative and positive numbers, the lowest value will be the most negative number. -1,560 is less than -156.

Question No.	Answer	Detailed Explanation
11	C, D	Both the options (C) and (D) are correct because their first number is a decimal that is greater than or equal to 1 but less than 10. They are then multiplied by a power of 10. One is multiplied by a positive power of 10, making it a large number and the second one is multiplied by a negative power, making it a really small number. Both are in correct form.
12	2.347×10^9	The number in the first box has to be a decimal number that is greater than or equal to 1, but less than 10. In this case it will be 2.347. Next, we will take it 10 times itself to an exponent which will move our decimal until it becomes our number in standard form. So in this case we have to take 10 times itself 9 times, so our exponent will be 9. Our final answer will be 2.347×10^9
13	6.87×10^{-5}	This time we are changing a really small number into scientific notation. The same principles apply. First we move our decimal to make it a number greater than one, but less than 10. In this case it becomes 6.87. Next we figure out the power of ten we need to move the decimal over that far. Since we are moving the decimal in the opposite direction, our power of ten will be negative this time. Here we are moving it 5 places, so it is 10^{-5}. Combine the two parts and you get 6.87×10^{-5}.

Lesson 4: Solving Problems Involving Scientific Notation

Question No.	Answer	Detailed Explanation
1	B	3.7×10^7 people ÷ 1.6×10^5 square miles =(3.7 ÷ 1.6) x 10^{7-5} = 2.3125×10^2 people/square mile = 231.25 people/ sq. mi. 240 is the best estimate.
2	C	6×10^7 kilometers ÷ 3×10^5 kilometers per second = (6÷3) x $10^{(7-5)}$ = 2×10^2(kilometers ÷ kilometers/seconds) = 2×10^2 seconds = 200 seconds
3	A	$(4 \times 10^6) \times (2 \times 10^3) = 8 \times 10^{(6+3)} = 8 \times 10^9$
4	D	$(2 \times 10^{-3}) \times (3 \times 10^5) = (2x3) \times 10^{(-3+5)} = 6 \times 10^2$
5	B	$(2.4 \times 10^3) / (6 \times 10) = 0.4 \times 10^2 = 4x10 = 40$
6	A	$(5x10^5) \times (9x10^{-3})=(45x10^{5-3})=45x10^2=4.5 \times 10^3$ Therefore, 4.5×10^4 is not equal to 4.5×10^3, so 4.5×10^4 is the correct answer. Options (B), (C) and (D) are different ways of expressing the same number.
7	A	$(5 \times 10^7) \div (10 \times 10^2) = 0.5 \times 10^{(7-2)} = 0.5 \times 10^5 = 5 \times 10^4$
8	C	$.00004567 \times .00001234 = (4.567 \times 10^{-5}) \times (1.234 \times 10^{-5}) = (4.567 \times 1.234) \times 10^{-5+(-5)} = 5.635678 \times 10^{-10}$ $5.636x10^{-10}$ is the correct answer.
9	D	$(50.67 \times 10^4) \times (12.9 \times 10^3) = 653.643 \times 10^{(4+3)} = 653.643 \times 10^7$ = 6,536,430,000 6,536,430,000 is the correct answer.
10	A	$(1.298 \times 10^4) \div (3.97 \times 10^2) = 0.326952 \times 10^2 = 32.7$ 32.7 is the best answer.
11	A, C	$(4.0 \times 10^3)(5.0 \times 10^5) = (4 \times 5) \times (10^{3+5}) = 20 \times 10^8 = 2.0 \times 10^9$. Therefore option (A) is correct. $(4.5 \times 10^5) / (9.0 \times 10^9) = (45 \times 10^4)/(9 \times 10^9) = (45/9) \times 10^{4-9} = 5.0 \times 10^{-5}$. This is not equal to 2.0×10^4. Therefore, option (B) is incorrect. $(2.1 \times 10^5) + (2.7 \times 10^5) = (2.1 + 2.7) \times 10^5 = 4.8 \times 10^5$. Therefore option (C) is correct. $(3.1 \times 10^5) - (2.7 \times 10^2) = (3.1 \times 10^3 - 2.7) \times 10^2 = (3100 - 2.7) \times 10^2 = (3097.3) \times 10^2 = 3.097 \times 10^5$. This is not equal to 0.4×10^3. Therefore option (D) is incorrect.

Question No.	Answer	Detailed Explanation
12	C, D	In these cases, you can compare numbers by looking at their exponents. A larger exponent means a larger number. To order numbers from largest to smallest, arrange them by their exponents from highest to lowest. For negative exponents, the number with the smaller absolute value of the exponent is larger.
13	5.946×10^4	When multiplying numbers in scientific notation, first multiply the decimal parts. Here, the product of 6 and 9.91 results in 59.46. Then add the exponents: 3 + 0 = 3. So, we have the product as 59.46×10^3. Since 59.46 is greater than 10, move the decimal one place left to get 5.946, and increase the exponent by 1 to 10^4. So, the final answer is 5.946×10^4.

Lesson 5: Compare Proportions

Question No.	Answer	Detailed Explanation
1	C	Unit rate means cost per unit (one). "Per" means to divide; so we divide the total cost by the number of units. $1440 ÷ 12 = $120 per unit
2	B	$2.90 ÷ 10 = $0.29 per piece $6.20 ÷ 20 = $0.31 per piece Therefore, Fruity is $0.02 more per piece than Big Bubbles.
3	A	Slope is defined as change in vertical height per unit change in horizontal distance. 9 ft ÷ 54 ft = $\frac{1}{6}$; 12 ft ÷ 84 ft = $\frac{1}{7}$ Since $\frac{1}{6} > \frac{1}{7}$, the first slope is steeper.
4	B	4 ft ÷ 12 ft = $\frac{1}{3}$ which is the largest ratio so the steepest slope.
5	A	The unit rate can also be called the average or mean, but not the mode, median, nor the frequency.
6	A	Unit cost is calculated by dividing the total cost by the amount of items. This statement is true. Therefore, "Unit cost is calculated by dividing the amount of items by the total cost." must be a false statement. (This would represent the number of items you get per dollar paid.)
7	D	1 qt = 32 fl oz 1 gallon = 128 fl oz 5 gallons = 640 fl oz $\frac{\$3.99}{24}$ fl oz = $0.16625 per fl oz $\frac{\$4.80}{32}$ fl oz = $0.15 per fl oz $\frac{\$11.00}{128}$ fl oz = $0.0859375 per fl oz $\frac{\$49.60}{640}$ fl oz = $0.0775 per fl oz--- So, the five gallon tub offers the best price
8	D	$\frac{5 \text{ ft}}{30 \text{ ft}} = \frac{1}{6}$ which is the least steep slope
9	B	6 tee shirt for $90 $90 ÷ 6 = $15 per tee shirt, which is the most expensive of the four options.

Question No.	Answer	Detailed Explanation
10	A	15 minutes ÷ 5 laps = 3 min/lap which is the fastest time
11	B	\$2.00 for a 6 pack-----\$0.33 per can \$6.00 for a 24 pack-----\$0.25 per can is the lowest unit cost \$3.60 for a 12 pack------\$0.30 per can \$10.00 for a 36 pack------\$0.28 per can
12	B	Plain wafers have 20 per pack and sell for \$2.40-------That is \$0.12 per wafer. Sugar-free wafers have 30 pieces per pack and sell for \$6.30-----That is \$0.21 per wafer. ---------\$0.09 more per unit than the plain wafers
13	A	36 ÷ 40 = 90% which is the highest score of the four practice tests.
14	D	\$30 ÷ 40 tickets = \$0.75 per ticket
15	C	Ski Slope A has a vertical rise of 4 feet and a horizontal run of 16 feet---$\frac{1}{4}$ Ski Slope B has a vertical rise of 3 feet and a horizontal run of 12 feet---$\frac{1}{4}$ Ski Slope C has a vertical rise of 3 feet and a horizontal run of 9 feet---$\frac{1}{3}$ has the steepest slope. Ski Slope D has a vertical rise of 5 feet and a horizontal run of 25 feet----$\frac{1}{5}$
16	y=14	To solve this proportion, use cross products. Multiply the numerator of one ratio by the denominator of the other, and vice versa. In this case we would have 7x4=2y. So 28=2y. Next you want to get the variable alone, so you have to use inverse operations to move the 2. In this case we will divide both sides by 2. Thus leaving us with y=14.

Question No.	Answer	Detailed Explanation
17	y=25	To solve this proportion you will have to use cross products. You will multiply the numerator of one to the denominator of the other and the denominator of the first to the numerator of the second. In this case we will multiply 20 x 20=16y. 400=16y. Then we have to use inverse operations to get y by itself. In this case we will divide both sides by 16. We will end up with y=25.
18	Lisa can buy = 8 kiwi fruit	You will set up the proportion so that it is the number of kiwi over the cost on both sides, making sure to have a variable where Lisa's number of kiwi fruit would go since that is what we are solving for. It should look like $\frac{32}{16} = \frac{X}{4}$. Now we have a proportion to solve. Cross multiply so you end up with (4)32=16x. 128=16x. Next, we will use our inverse operations to get x alone. When we divide both sides by 16 we end up with x=8. So Lisa can purchase 8 kiwi fruits.

Lesson 6: Understanding Slope

Question No.	Answer	Detailed Explanation
1	D	Only the last statement, "Slope is determined by dividing the vertical distance between two points by the corresponding horizontal distance," is true.
2	D	The slope of the line passing through these two points is $\frac{4 - 2}{-1 - 1} = \frac{2}{-2} = -1$. The only equation with a slope of - 1 is y = - x + 3.
3	D	The slope of y = 4x + 3 is 4. The slope of y = 4x - 2 is 4.
4	D	$y = \frac{9}{5}x - 3$ has a slope of $\frac{9}{5}$ which is the greatest.
5	A	$\frac{1}{8}$ is the smallest slope. $y = \frac{1}{8}x + 7$ is the correct answer.
6	A	If we know the y-intercept and the slope, we can write the equation of a straight line. If we know the y-intercept, we know b in the slope-intercept formula. The y-intercept together with another point or the x-intercept make it possible to determine the slope of the line. Direction would not give enough information
7	C	Slope = rise/run (vertical change/horizontal change) Slope $= \frac{30}{500} = \frac{3}{50}$
8	C	y = mx + b m = slope In this case, we have $y = (\frac{3}{2})x - 9$ so $m = \frac{3}{2}$
9	A	We know the y-intercept is (0, 5) and can determine the slope is $\frac{10}{2} = 5$, giving us the equation of the line: y = 5x + 5. We can check by substituting the coordinates for x and solve for y. (0,5): y = 5x + 5 = 5(0) + 5 = 5 = y (2,15): y = 5x + 5 = 5(2) + 5 = 10 + 5 = 15 = y

Question No.	Answer	Detailed Explanation
10	A	$m = \frac{10 - 6}{3 - 1} = \frac{4}{2} = 2$ The equation where the slope equals 2 is y = 2x - 9.
11	A	The slope, $m = (\frac{10 - 6}{5 - 3})$ is 2. The equation where the slope equals 2 is y = 2x - 12
12	B	The equation y = -5x - 4 has -5 as its slope, and the slope of y = -5x - 11 is also -5.
13	B	(5 - 3) ÷ (5 - 3) = 1 The slope is 1.
14	A	The slopes of these lines are 4, -3, and 3 respectively. Of these, 4 would be the steepest slope.
15	C	(0,-3) and (-3,0) The slope of the line passing through these two points is (0 - - 3) ÷ (- 3 - 0) = 3 ÷ - 3 = - 1. We know that the y-intercept is at (0, - 3) so the line is y = -x - 3.
16	The slope is $= \frac{13}{12}$	To find the slope between 2 ordered pairs you use the formula $\frac{y_2 - y_1}{x_2 - x_1}$. $\frac{8-(-5)}{0-(-12)} = \frac{13}{12}$ So the slope between the two points is $\frac{13}{12}$
17	The slope is $= \frac{1}{9}$	To find the slope between 2 ordered pairs you use the formula $\frac{y_2 - y_1}{x_2 - x_1}$. $\frac{-2-(-3)}{12-3} = \frac{1}{9}$ So the slope between the two points is $\frac{1}{9}$.
18	$\frac{-9}{4}$	To find the slope between 2 ordered pairs you use the formula $\frac{y_2 - y_1}{x_2 - x_1}$. So the slope between the two points is $\frac{-7 - 2}{5 - 1}$ $= \frac{-9}{4}$

Lesson 7: Solving Linear Equations

Question No.	Answer	Detailed Explanation
1	A	Consecutive odd integers lie 2 units apart on the number line; so let the numbers be represented by n and n+2. Their sum is 44; so n + n + 2 = 44. 2n +2 = 44 2n = 44 - 2 2n = 42 n = 21; so n + 2 = 23
2	A	Cumulative GPA = $\frac{2.1 + 2.9 + 3.1 + n}{4}$ $3.0 = \frac{8.1 + n}{4}$ 12.0 = 8.1 + n 3.9 = n
3	C	Let m = miles/day for remaining 3 days m = (original miles - miles already traveled) / 3 $m = \frac{1924 - 490}{3} = \frac{1434}{3} = 478$ miles per day
4	C	3x + 5 = 29 3x = 29 - 5 3x = 24 x = 8
5	D	7 - 2x = 13 - 2x Subtract 13 from both sides: 7 - 13 = -2x + 2x - 6 = 0 Since this is false, no solution exists.
6	B	6x + 1 = 4x - 3 6x - 4x = - 3 - 1 2x = - 4 $x = -\frac{4}{2}$ x = -2
7	D	2x + 6 + 1 = 7 + 2x 2x + 7 = 7 + 2x Since both sides of the equation are identical, any real solution satisfies the equation.

Question No.	Answer	Detailed Explanation
8	D	8x - 1 = 8x -1 = 0 Since this is a false statement, there is no solution to this equation.
9	D	2x + 5x - 9 = 8x - x - 3 - 6 7x - 9 = 7x - 9 All real values for x are solutions.
10	D	-y + 7y -54 = 0 6y - 54 = 0 6y = 54 y = 9
11	A, C & D	To solve these equations, you will use inverse operations. Depending on whether it is a one-step or two-step equation, you might use one or two operations. The first one is a one step equation. Since it is a multiplication problem, you will use division to undo it. In this case we are going to divide both sides by -5. When we do this, we get m = -5. The second one is also a one step equation. We divide both sides by -10. When we do this, we get c = 8. The third one is a one step equation. Since it is a addition problem, to undo the addition, we will subtract. When we subtract a -7 from both sides, we end up with g = -5. The last one is a two step equation. In this case we have to subtract 20 from both sides first. We end up with 12m = -60. Now we have to undo the multiplication. To do this we will divide both sides by 12. When we do this, we will end up with m = -5.
12	-8	To solve this 2-step equation, first subtract 20 from both sides to undo the addition. Making it now 5x=-40. Now the second step is to undo the multiplication. To undo it, we will divide both sides by 5. Ending up with x=-8.
13	missing number = 13	Since you are already given the value of x, the first thing you want to do is substitute the value in. When you do this you end up with 7+(-9)=(-15)+?. Next combine your terms so you end up with -2 = -15 + ?. Next you will move the -15 to the other side. To do this you will add 15 to both sides. You end up with the unknown being 13.

Lesson 8: Solve Linear Equations with Rational Numbers

Question No.	Answer	Detailed Explanation
1	C	$\frac{7}{14} = n + (\frac{7}{14})(n)$ $\frac{1}{2} = n + (\frac{1}{2})n$; Multiply by 2 to eliminate denominators. $1 = 2n + n$ $1 = 3n$ $\frac{1}{3} = n$
2	B	$2(2x - 7) = 14$ $4x - 14 = 14$ $4x = 14 + 14$ $4x = 28$ $x = 7$
3	D	$6x - (2x + 5) = 11$ $6x - 2x - 5 = 11$ $4x - 5 = 11$ $4x = 11 + 5$ $4x = 16$ $x = 4$
4	B	$4x + 2(x - 3) = 0$ $4x + 2x - 6 = 0$ $6x - 6 = 0$ $6x = 6$ $x = 6 \div 6$ $x = 1$
5	A	$3y - 7(y + 5) = y - 35$ $3y - 7y - 35 = y - 35$ $- 4y - 35 = y - 35$ $- 35 = y - 35 + 4y$ $- 35 + 35 = y + 4y$ $0 = 5y$ $0 = y$
6	A	$2(x-5) = \frac{1}{2}(6x + 4)$ $2x - 10 = 3x + 2$ $-12 = x$

Question No.	Answer	Detailed Explanation
7	D	$3x + 2 + x = \frac{1}{3}(12x + 6)$ $3x + x + 2 = 4x + 2$ $4x + 2 = 4x + 2$ $0 = 0$ Thus, all real values of x are correct solutions.
8	C	$\frac{1}{2}x + \frac{2}{3}x + 5 = \frac{5}{2}x + 6$ $6(\frac{1}{2}x + \frac{2}{3}x + 5 = \frac{5}{2}x + 6)$ $3x + 4x + 30 = 15x + 36$ $-8x = 6$ $x = -\frac{6}{8} = -\frac{3}{4}$
9	B	$2(2x + 3) = 3(2x + 5)$ $4x + 6 = 6x + 15$ $-2x + 6 = 15$ $-2x = 9$ $x = -\frac{9}{2}$
10	B	$.64x - .15x + .08 = .09x$ $.49x + .08 = .09x$ $.08 = -.04x$ $-.2 = x$
11	A, C	A and C are correct options. To start with the first one, you need to get the variable alone, so you add $\frac{2}{5}$ to both sides. When you do this you are left with $w=\frac{10}{5}$, which simplifies to w=2. The third one is still dealing with rational numbers, but this time in the form of decimals. First step is to get the variables on one side and the numbers on the other. So, first you will add 1.2 to both sides. You get $0.4x = 0.15x + 2$. Next move the variables to one side. You will subtract 0.15x from both sides. This will leave you with $0.25x = 2$. Last step is to divide both sides by 0.25 (or multiply by 4), and you are left with $x = 8$.

Question No.	Answer	Detailed Explanation
12	B	The equation starts off like . $\frac{3}{4} + \frac{1}{2}(m+\frac{1}{4}) = \frac{19}{16}$ The first step is to eliminate the ¾. from the left-hand side of the equation. To do this you will subtract ¾ from both sides. On the right hand side you find a common denominator before you subtract. Then you will need to get rid of the ½. To do this you can either divide both sides by ½ or multiply by 2. Lastly, you need to get rid of the ¼. To do this you will subtract ¼ from each side. This will finally reveal your answer.
13	$n = \frac{1}{3}$	$\frac{8}{16} = n + \frac{8}{16}n$ Multiplying both sides by 2, we get, $2 \times \frac{8}{16} == 2\ (n + \frac{8}{16}n)$ 1 = 2n+n 3n = 1 Therefore, $n = \frac{1}{3}$

Lesson 9: Solutions to Systems of Equations

Question No.	Answer	Detailed Explanation
1	D	$x - 5 = 2x + 1$ $- 5 - 1 = 2x - x$ $-6 = x$ Then $y = x - 5$ $y = - 6 - 5$ $y = - 11$
2	C	These lines have the same slope, but different y-intercepts, so they do not intersect. There is no solution. The lines are parallel.
3	D	$y = 2(2 - 3x) = - 6x + 4$ $y = -3(2x + 3) = - 6x - 9$ There is no solution because for any value for x, we have 2 different values for y. This is inconsistent.
4	B	$\frac{x}{2} + \frac{y}{3} = 2$ $3x - 2y = 48$ multiplying the first equation by 6 changes the left side to $3x + 2y = 12$, and adding it to the second equation eliminates y. $x = 10$ $3(10) - 2y = 48$ $30 - 2y = 48$ $-2y = 48 - 30 = 18$ $y = -9$
5	B	$y = 2x - 6$ $y = -2x + 6$ $2x - 6 = - 2x + 6$ $2x + 2x = 6 + 6$ $4x = 12$ $x = 3$ $y = 2(3) -6 = 0$
6	A	Adding the two equations, we get $4x = 4$; $x = 1$ Then substituting $x = 1$ into $2x + 3y = 14$, we get $2(1) + 3y = 14$ $3y = 14 - 2$ $3y = 12$ $y = 4$

Question No.	Answer	Detailed Explanation
7	D	Solving the second equation for x, we get x = 13 + 2y which is identical to the first equation. Therefore, there are infinitely many points (x, y) that satisfy the system.
8	B	Solving x + y = 9 for y, we get y = - x + 9. Then 3x - 7 = - x + 9 3x + x = 9 + 7 4x = 16 x = 4 y = 5
9	C	Examining these equations we find an inconsistency. 2x + 5y CANNOT be equal to two different quantities. Therefore, there is no solution.
10	A	Adding the two equations, we get 14y = 28. y = 2 Substituting into 4x + 7y = 2, we get 4x + 14 = 2 4x = -12 x = - 3
11	B	Let x = the number of $1 bills that he has. Let y = the number of $5 bills that he has. x + 5y = 49 (Eq. 1) x = y + 1 x - y = 1,. Multiplying this equation by 5, we get 5x - 5y = 5 (Eq. 2) Adding Eq. 1 and Eq. 2, 6x = 54 x = 9 ($1 bills) Substituting x = 9 in x - y = 1, we get, y = 8 ($5 bills)
12	D	Let x = Cole's age now and y = Anya's age now. y = x + 3. This can be rewritten as y - x = 3 x + 11 = 2y 2y - x = 11 Eqn. 1 Multiplying the equation, y - x = 3 by -1, we get -y + x = -3 Eqn. 2 Adding Eqn. 1 and Eqn. 2, we get y = 8. Substituting y = 8 in, y - x = 3, we get x = 5. y = 8 years old x = 5 years old

Lesson 10: Solving Systems of Equations

Question No.	Answer	Detailed Explanation
1	A	$13x + 3y = 15$ $4x + y = 5$ Multiply the bottom equation by -3 to get $-12x - 3y = -15$. Then add this to the first equation. $x = 0$ $x = 0$ Substitute into $y = 5 - 4x$ and $y = 5 - 0$ $y = 5$
2	A	$y = 2x + 5$ $y = 3x - 7$ $2x + 5 = 3x - 7$ $5 = 3x - 7 - 2x$ $5 + 7 = 3x - 2x$ $12 = x$ $y = 2(12) + 5 = 24 + 5 = 29$
3	A	Adding the two equations, we get $4x = 4$; $x = 1$ Then substituting into $2x + 3y = 14$, we get $2 + 3y = 14$ $3y = 14 - 2$ $3y = 12$ $y = 4$
4	D	Solving the second equation for x, we get $x = 13 + 2y$ which is identical to the first equation. Therefore, there are infinitely many points (x, y) that satisfy the system.
5	C	Examining these equations, we find an inconsistency. $2x + 5y$ cannot be equal to two different quantities. Therefore, there is no solution.
6	A	Adding the two equations, we get $14y = 28$. $y = 2$ Substituting into $4x + 7y = 2$, we get $4x + 14 = 2$ $4x = -12$ $x = -3$

Question No.	Answer	Detailed Explanation
7	B	$y + 3x = 11$ $y - 2x = 1$ Multiplying the equation $(y + 3x = 11)$ by 2 and the equation $(y - 2x = 1)$ by 3, we get $2y + 6x = 22$ Eq. 1 $3y - 6x = 3$ Eq. 2 Adding Eq. 1 and Eq. 2, $5y = 25$ $y = 5$; Substituting $y = 5$ in $y - 2x = 1$ $5 - 2x = 1$; $-2x = 1-5 = -4$ $2x = 4$ $x = 2$
8	D	$2x + 4y = 14$ $x + 2y = 7$ Dividing the first equation by 2, we obtain the second equation. Therefore, there are an infinite number of solutions.
9	B	$3(y - 2x) = 9$ $x - 4 = 0$ $x = 4$ $3y - 6x = 9$ $3y - 6(4) = 9$ $3y - 24 = 9$ $3y = 9 + 24$ $3y = 33$ $y = 11$
10	D	$10x = -5((3x - 7)+2)$ $10x = -5(3x - 5)$ $10x = -15x+ 25$ $25x = 25$ $x = 1$ $y = 3(1) - 7$ $y = -4$

Question No.	Answer	Detailed Explanation
11	A, C	These systems of equations [options (A) and (C)] will intersect at exactly one point. If you solve the first one, it will intersect at (-4, -2). The other one will intersect at (3, -2). System of equations in the option (B) represents pair of parallel lines as they have same slope but different y - intercept. System of equations in option (D) represents the same line as they have same slope and same y - intercept. So this system has infinite solutions. (Note that the equation 4y = 2x - 4 is essentially same as -x + 2y = -2, when you take out the common factor 2 and rearrange the terms).
12		After solving all the above system of equations, we see that: (-4,-3) satisfies the first system of equations; (-1,2) satisfies the second system of equations; (1,4) satisfies the third system of equations; (-3,-2) satisfies the fourth system of equations.
13	B, C & D	Options B, C and D are correct answers. These all have no solution because the equations in each of these systems will make parallel lines. Since the lines are parallel, they will not intersect, resulting in no solution.
14	No Solution, One Solution & Infinite Solution.	The first one has no solution because if you rearrange these into slope intercept form, you will see that they both have the same slope but different y - intercept. If they have the same slope and different y-intercept, they will be parallel and will not intersect. The second one has one solution because if you solve the system you will end up with one value for x and y, giving you the ordered pair where they intersect. The other equation is also rearranged and that both equations represent the same line, leading to infinite solutions.

Lesson 11: Systems of Equations in Real-World Problems

Question No.	Answer	Detailed Explanation
1	C	Jorge: $y = 50 + 1x$ Jillian: $y = 72 + .12x$ $50 + 1x = 72 + .12x$ $.88x = 22$ $x = 25$
2	B	Now: Tom = x; Mr. Stevens = y and $y = x + 63$ In 3 Yrs: Tom = x + 3; Mr. Stevens = y + 3 and $y + 3 = 4(x + 3)$ Simplify $y + 3 = 4(x + 3)$ $y = 4x + 12 - 3$ $y = 4x + 9$ and then our system of equations is $y = x + 63$ and $y = 4x + 9$ Multiply top equation by - 4 and $-4y = -4x - 252$; Add $y = 4x + 9$ $-3y = -243$ $y = 81$ Substitute into $y = x + 63$ $81 = x + 63$ and $18 = x$ which is Tom's age The correct answer is 18 years.
3	C	There are a total of 50 textbooks and workbooks in the box; so $t + w = 50$. Each textbook weighs 2 lb and each workbook weighs 0.5 lb and the total weight is 55 lb; so $2t + 0.5w = 55$.
4	D	Total cost for Plumber A is $40x + 50$ and for Plumber B is $35x + 75$.
5	A	$10t + 6d = 19.50$ $7t + 5d = 14.25$ correctly expresses the relative costs as described.

Question No.	Answer	Detailed Explanation
6	C	Let C = money Cindy has. Let M = money Mindy has. C = 25 + 12w where w is the weeks worked. M = 55 + 7w where w is the weeks worked. 25 +12w = 55 + 7w ; 12w - 7w = 55 - 25 5w = 30 ; w = 6 weeks ; C = 25 + 12(6) = $97.00 M = 55 + 7(6) = $97.00
7	C	Let x = calendars and y = candy bars x + y = 1100 ; 1.50x = 1.25y Multiply the equation x + y = 1100 by 1.50 1.50x + 1.50y = 1.50(1100) 1.50x + 1.50y = 1650 Eq.1 1.50x = 1.25y can be rewritten as 1.50x-1.25y = 0 Eq. 2; Subtracting Eq. 2 from Eq. 1, we get 2.75y=1650 y= 600 candy bars sold Substituting y = 600 in x + y = 1100, we get x = 1100 - y = 500 calendars
8	D	Let x = Cole's age now and y = Anya's age now. y = x + 3. This can be rewritten as y - x = 3 x + 11 = 2y 2y - x = 11 Eqn. 1 Multiplying the equation y - x = 3 by -1, we get -y + x = -3 (Eqn. 2). Adding Eqn. 1 and Eqn. 2, we get y = 8. Substituting y = 8 in, y - x = 3, we get x = 5. y = 8 years old and x = 5 years old
9	D	Tom: y = 15 + 3x; Sister: y = 7 + 5x; 15 + 3x = 7 + 5x; 8 = 2x ; x = 4
10	B	Correct Answer: B Let x denote the day. Let y be the number of miles Lucia runs on the xth day. Let z be the number of miles Jack runs on the xth day; y = 2 + (0.25)(x-1) = 2 + 0.25x - 0.25 = 1.75 + 0.25x; So, for Lucia, we have the equation : y = 1.75 + 0.25x Eqn. 1; z = 0.5 + 0.5(x - 1) = 0.5 + 0.5x - 0.5 = 0.5x. So, for Jack, we have the equation: z = 0.5x (Eqn. 2). We want to know the value of x for which y = z; Equating Eqn. 1 and Eqn 2, we get, 0.5x = 1.75 + 0.25x; 0.5x - 0.25x = 1.75 ; 0.25x = 1.75; x = $\frac{1.75}{0.25}$ = 7 So, they run the same distance on the 7th day.

Question No.	Answer	Detailed Explanation
11	B, C	1500 people enter the fair which includes children and adults and can be represented by c+a=1500. The admission fee is \$3 for children and \$5 for adults. So 3c+5a=5740 would represent the total fair collected. Hence option B is the correct answer. Solving these equations will give us a=620 and c=880. Therefore, option C is also the correct answer.
12		In solving the equations, it is enough to calculate the value of one variable. (1) c = cost of 1 bag of chips, p = cost of 1 bag of pretzels. Then the equations are 5c + 9p = 16.95 and 10c + 10p = 25.50. Dividing the 2nd equation by 2, we get 5c + 5p = 12.75. 5c + 9p = 16.95 5c + 5p = 12.75 Subtracting, we get 4p = 4.20 or p = 1.05. Therefore option (B) is the correct answer for this system. (2) n = number of nickels, q = number of quarters. Then the equations are n + q = 49 and 0.05n + 0.25q = 8.65. Dividing the 2nd equation by 0.05, we get n + 5q = 173. n + 5q = 173 n + q = 49 Subtracting, we get 4q = 124 or q = 31. Therefore option (D) is the correct answer for this system. Alternate Method : Since the total number of coins is 49, you can easily guess that option (D) is the correct choice (because 18 + 31 = 49). (3) d = cost of 1 dog and s = cost of 1 soft drink. Then the equations are d + 2s = 4.95 and 3d + s = 6.85. Multiplying the 2nd equation by 2, we get 6d + 2s = 13.70 6d + 2s = 13.70 d + 2s = 4.95 Subtracting, we get 5d = 8.75 or d = 1.75. Therefore option (C) is the correct answer for this system. (4) Remaining option (A) is the correct answer to problem 4. Alternately, since the total number of animals is 34, you can easily guess this is the correct option. Alternate Method : So, if you guess solutions to problem 2 and problem 4 at the beginning. you need to solve only one set of equations, say for problem 1.

Chapter 3: Functions

Lesson 1: Functions

1. Which of the following is NOT a function?

Ⓐ {(2, 3), (4, 7), (8, 6)}
Ⓑ {(2, 2), (4, 4), (8, 8)}
Ⓒ {(2, 3), (4, 3), (8, 3)}
Ⓓ {(2, 3), (2, 7), (8, 6)}

2. Which of the following tables shows that y is a function of x?

Ⓐ

x	y
1	4
1	7
4	7

Ⓑ

x	y
1	7
3	8
4	7

Ⓒ

x	y
3	2
3	7
4	8

Ⓓ

x	y
1	7
4	7
4	9

3. If y is a function of x, which of the following CANNOT be true?

Ⓐ A particular x value is associated with two different y values.
Ⓑ Two different x values are associated with the same y value.
Ⓒ Every x value is associated with the same y value.
Ⓓ Every x value is associated with a different y value.

4. **Each of the following graphs consists of two points. Which graph could NOT represent a function?**

Ⓐ 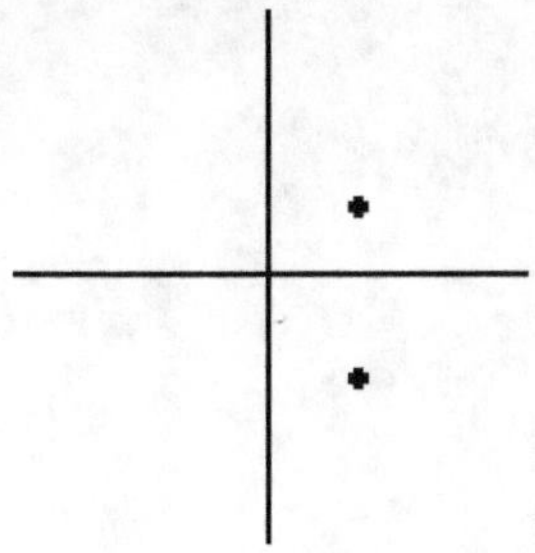

Ⓑ 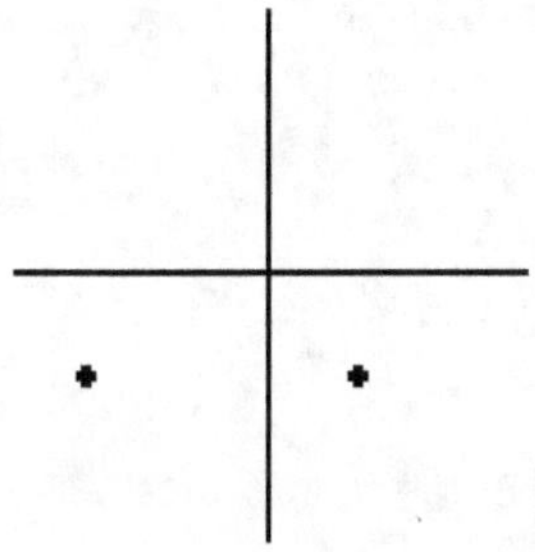

Ⓒ 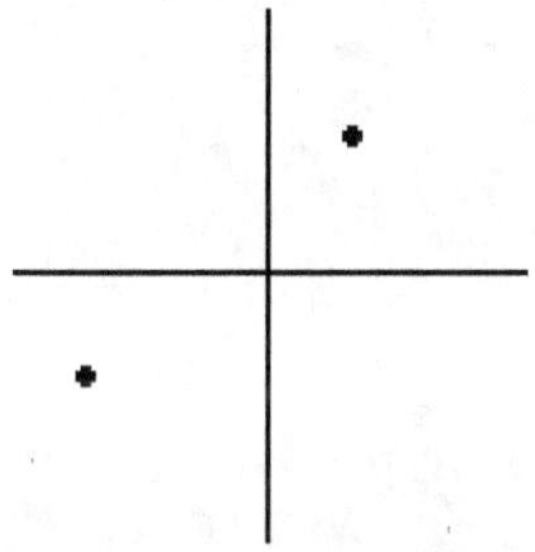

Ⓓ 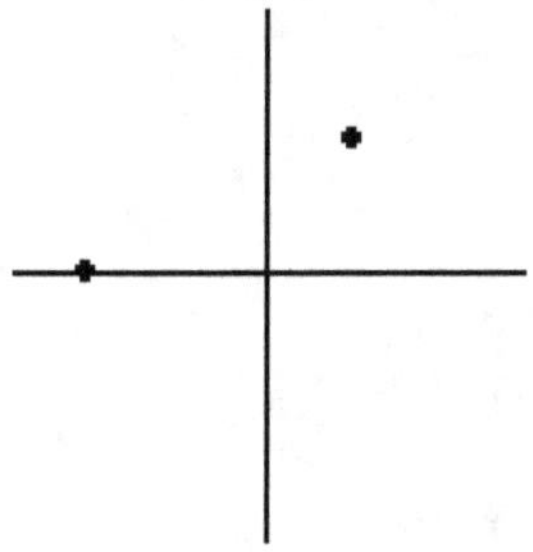

5. **Which of the following could NOT be the graph of a function?**

Ⓐ 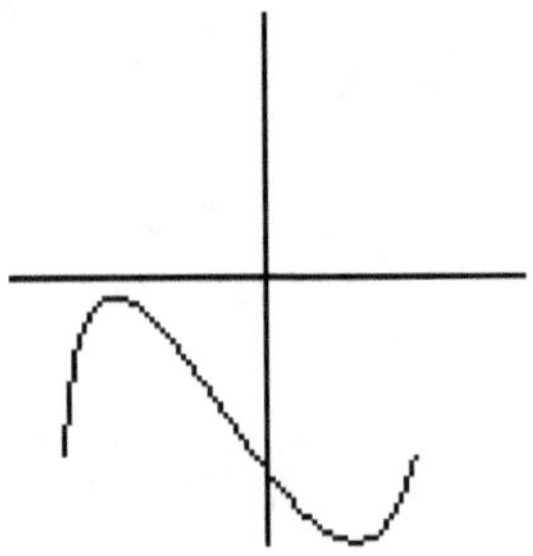

Ⓑ 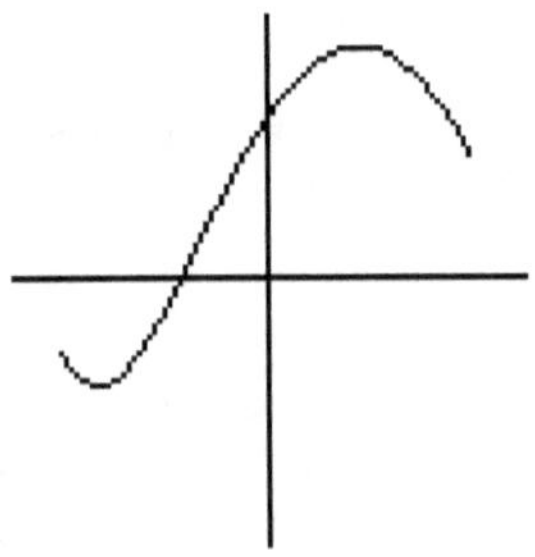

Ⓒ 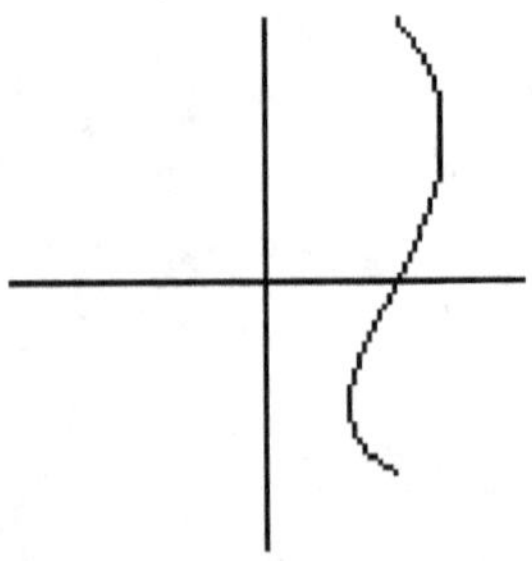

Ⓓ 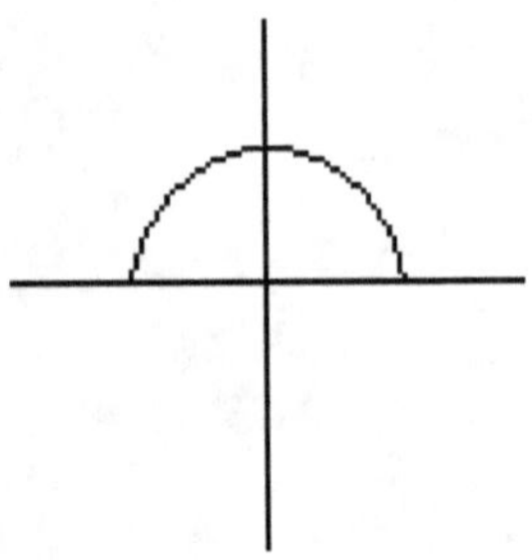

6. **Which of the following is NOT a function?**

Ⓐ {(0, 0), (2, 2), (4, 4)}
Ⓑ {(0, 4), (2, 4), (4, 4)}
Ⓒ {(0, 0), (2, 0), (4, 0)}
Ⓓ {(0, 0), (0, 2), (0, 4)}

7. Which of the following is true of the graph of any non-constant function?

Ⓐ A line drawn parallel to the x-axis will never cross the graph.
Ⓑ A line drawn perpendicular to the x-axis will never cross the graph.
Ⓒ A line drawn through the graph parallel to the x-axis will cross the graph one and only one time.
Ⓓ A line drawn through the graph perpendicular to the x-axis will cross the graph one and only one time.

8. The given set represents a function:
{(0,1), (1,1), (2,1)}
If the ordered pair ____ was added to the set, it would no longer be a function.

Ⓐ (3,1)
Ⓑ (3,2)
Ⓒ (3,3)
Ⓓ (2,3)

9. In order for this set, {(6,5), (5, 4), (4,3)}, to remain a function, which of the following ordered pairs COULD be added to it?

Ⓐ (6,6)
Ⓑ (5,5)
Ⓒ (4,4)
Ⓓ (3,3)

10. Which of the following sets of ordered pairs represents a function?

Ⓐ [(0,1), (0,2), (1,3), (1,4)]
Ⓑ [(1,1), (1,2), (1,3), (1,4)]
Ⓒ [(2,5), (2,6), (4,7), (5,7)]
Ⓓ [(-7,10), (7,10), (8,9), (9,10)]

11. Select the sets of ordered pairs that represent a function.

There can be more than one correct option. Select all the correct options.

Ⓐ {(1,1),(2,2),(3,3),(4,4)}
Ⓑ {(1,-2),(-2,0),(-1,2),(1,3)}
Ⓒ {(-2,3),(0,1),(2,-1),(3,-2)}
Ⓓ {(-1,7),(0,-3),(1,10),(0,7)}

12. Write the function that would go with the table.

x	2	3	4	5
y	-1	0	1	2

13. Select whether or not each example represents a function.

	Function	Not a Function
{(2,10),(2,20),(4,20),(6,30)}	○	○
{(4,10),(8,12),(4,11),(12,13)}	○	○
{(1,3),(2,4),(3,7),(4,13)}	○	○
y=7x-2	○	○

CHAPTER 3 → Lesson 2: Comparing Functions

1. A set of instructions says to subtract 5 from a number n and then double that result, calling the final result p. Which function rule represents this set of instructions?

Ⓐ p = 2(n – 5)
Ⓑ p = 2n – 5
Ⓒ n = 2(p – 5)
Ⓓ n = 2p – 5

2. Which of the following linear functions is represented by the (x, y) pairs shown in the table below?

x	y
-3	-1
1	7
4	13

Ⓐ y = x + 2
Ⓑ y = 2x + 5
Ⓒ y = 3x + 1
Ⓓ y = 4x + 3

3.

x	y
0	3
1	5
2	7

Three (x, y) pairs of a linear function are shown in the table above. Which of the following functions has the same slope as the function shown in the table?

Ⓐ y = 3x + 2
Ⓑ y = 2x -4
Ⓒ y = x + 3
Ⓓ y = x - 1

4. The graph of linear function A passes through the point (5, 6). The graph of linear function B passes through the point (6, 7). The two graphs intersect at the point (2, 5). Which of the following statements is true?

Ⓐ Function A has the greater slope.
Ⓑ Both functions have the same slope.
Ⓒ Function B has the greater slope.
Ⓓ No relationship between the slopes of the lines can be determined from this information.

5. If line M includes the points (-1, 4) and (7, 9) and line N includes the points (5, 2) and (-3, -3), which of the following describes the relationship between M and N?

Ⓐ They are perpendicular.
Ⓑ They intersect but are not perpendicular.
Ⓒ They are parallel.
Ⓓ Not enough information is provided.

6. If line R includes the points (-2, -2) and (6, 4) and line S includes the points(0,4) and (3,0) which of the following describes the relationship between R and S?

Ⓐ They are perpendicular.
Ⓑ They intersect but are not perpendicular.
Ⓒ They are parallel.
Ⓓ Not enough information is provided.

7.

x	y
-4	13
2	1
6	-7

Three points of a linear function are shown in the table above. What is the y-intercept of this function?

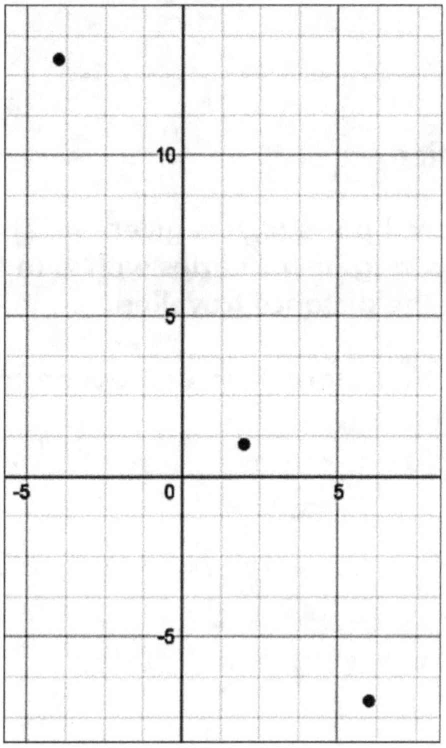

Ⓐ **(0, 5)**
Ⓑ **(0, 6)**
Ⓒ **(0, 7)**
Ⓓ **(0, 8)**

8. **Comparing the two linear functions, y = 2x + 7 and y = 7x + 2, we find that**

Ⓐ **y = 2x + 7 has the steeper slope.**
Ⓑ **The graphs of the functions will be parallel lines.**
Ⓒ **The graphs of the functions will intersect at the point (1, 9).**
Ⓓ **The graphs of the functions will be perpendicular lines.**

9. Which of the following is true of the graphs of the lines $y = 3x + 5$ and $y = -\frac{x}{3} + 6$?

Ⓐ The graphs will be two parallel straight lines.
Ⓑ The graphs will be two perpendicular straight lines.
Ⓒ Both functions will produce the same line.
Ⓓ There is not enough information to compare.

10. If line A, denoted by $y = x + 9$, and line B, denoted by $y = 5x - 3$ are graphed, which of the following statements is correct?

Ⓐ Line B has a steeper slope.
Ⓑ Both lines have the same slope.
Ⓒ Line A has the steeper slope.
Ⓓ There is not enough information to answer the question.

11. A function may be described in 3 ways : (1) Set of ordered pairs (x,y) is given or (2) An equation showing how y depends on x or (3) a table showing how y varies with x. In the problem given below, take x to be the time and y to be the distance travelled.

Select the ones that have the correct conclusion. Note that more than one option may be correct.

Ⓐ **CJ' s Speed:**

X	Y
0	0
2	13
4	26

Holly's Speed: $y=6x$ Conclusion: Holly is faster than CJ.

Ⓑ **Hank's Speed:** $y=\frac{7}{2}x$ **Smith's Speed:** {(0,0),(4,5),(8,10)} Conclusion: Hank is faster than Smith.

Ⓒ **Tyler's Speed:**

X	Y
0	0
4	6
8	12

Omar's Speed: $y=\frac{3}{2}x$ Conclusion: Tyler and Omar run the same speed.

12. **Put the correct inequality or equality sign between the rates of change of the two functions given below.**

Instruction : Take f to be the first function {(0, 0), (3, 180), (6, 360), (9, 540)} and g to be the second function y = 100x

{(0, 0), (3, 180), (6, 360), (9, 540)} [ ] y = 100x

13. **Match each function to whether the rate of change is positive or negative**

	Positive	Negative
$y=3x+4$	○	○
$y=\frac{1}{2}x-6$	○	○
$y=-7x+2$	○	○
$y=\frac{2}{3}x-1$	○	○

CHAPTER 3 → Lesson 3: Linear Functions

1. A linear function includes the ordered pairs (2, 5), (6, 7), and (k, 11). What is the value of k?

Ⓐ 8
Ⓑ 10
Ⓒ 12
Ⓓ 14

2. Which of the following functions is NOT linear?

Ⓐ $f(x) = x + 0.5$
Ⓑ $f(x) = -x + 0.5$
Ⓒ $f(x) = x^2 + 0.5$
Ⓓ $f(x) = 0.5x$

3. Which of the following functions is linear and includes the point (3, 0)?

Ⓐ $f(x) = 3/x$
Ⓑ $f(x) = x - 3$
Ⓒ $f(x) = 3$
Ⓓ $f(x) = 3x$

4. Four (x, y) pairs of a certain function are shown in the table below. Which of the following describes the function?

x	y
-3	1
-1	4
1	7
3	10

Ⓐ The function increases linearly.
Ⓑ The function decreases linearly.
Ⓒ The function is constant.
Ⓓ The function is not linear.

5. **Four (x, y) pairs of a certain function are shown in the table below. Which of the following statements describes the function correctly?**

x	y
0	3
1	4
2	7
3	12

Ⓐ **The function is linear because it does not include the point (0, 0).**
Ⓑ **The function is linear because it does not have the same slope between different pairs of points.**
Ⓒ **The function is nonlinear because it does not include the point (0, 0).**
Ⓓ **The function is nonlinear because it does not have the same slope between different pairs of points.**

6. **The graph of a linear function lies in the first and fourth quadrants. Which of the following CANNOT be true?**

Ⓐ **It is an increasing function.**
Ⓑ **It is a constant function.**
Ⓒ **It also lies in the second quadrant.**
Ⓓ **It also lies in the third quadrant.**

7. **A linear function includes the ordered pairs (0, 1), (3, 3), and (9, n). What is the value of n?**

Ⓐ **5**
Ⓑ **6**
Ⓒ **7**
Ⓓ **8**

8. **A linear function includes the ordered pairs (0, 3), (3, 9), and (9, n). What is the value of n?**

Ⓐ **20**
Ⓑ **14**
Ⓒ **21**
Ⓓ **22**

9. The graph of a linear function with a non-negative slope lies in the first and second quadrants. Which of the following CANNOT be true?

Ⓐ It is an increasing function.
Ⓑ It is a constant function.
Ⓒ It also lies in the third quadrant.
Ⓓ It also lies in the fourth quadrant.

10. Which function is represented by this table?

x	y
0	2
1	4
2	6
3	8

Ⓐ $y = 2x + 2$
Ⓑ $y = 3x - 4$
Ⓒ $y = 4x - 5$
Ⓓ $y = 6x - 8$

11. Which of the following functions are linear? Select all the correct answers.

Ⓐ 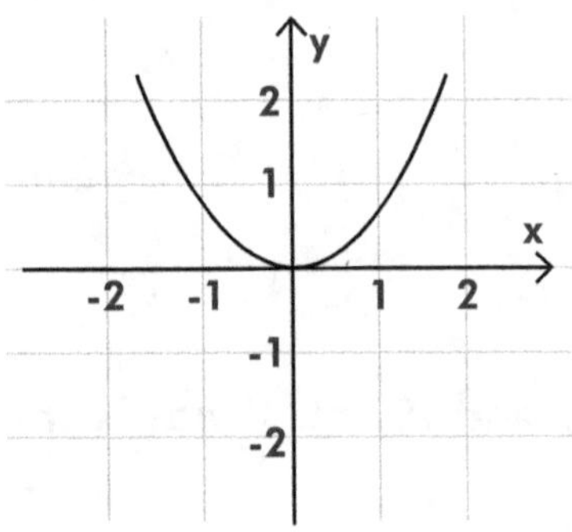

Ⓒ

x	y
-2	-4
-1	-1
0	2
1	5
2	8

difference of y-values

(-1) + 4 = 3,

2 + 1 = 3,

5 - 2 = 3,

8 - 5 = 3

Ⓑ $y = 5x+9$

Ⓓ $y=x^2+6$

12. **Write whether the equation represents a linear or non-linear function.**

$2x^2+3y=10$

13. **Match each function to linear or non-linear.**

<table>
<tr><th></th><th>Linear</th><th>Non-Linear</th></tr>
<tr><td>$y = x^2$</td><td>○</td><td>○</td></tr>
<tr><td>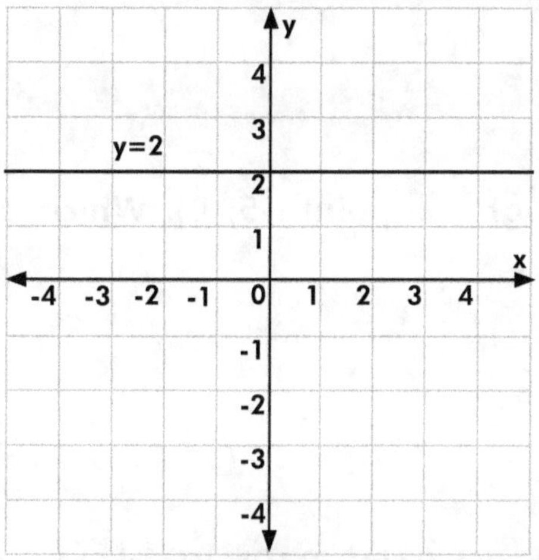</td><td>○</td><td>○</td></tr>
<tr><td><table><tr><th>x</th><th>y</th></tr><tr><td>1</td><td>1</td></tr><tr><td>2</td><td>8</td></tr><tr><td>3</td><td>27</td></tr><tr><td>4</td><td>64</td></tr><tr><td>5</td><td>125</td></tr><tr><td>6</td><td>216</td></tr><tr><td>7</td><td>?</td></tr></table></td><td>○</td><td>○</td></tr>
<tr><td>$y=-4x+12$</td><td>○</td><td>○</td></tr>
</table>

CHAPTER 3 → Lesson 4: Linear Function Models

1. If a graph includes the points (2, 5) and (8, 5), which of the following must be true?

   Ⓐ It is the graph of a linear function.
   Ⓑ It is the graph of an increasing function.
   Ⓒ It is not the graph of a function.
   Ⓓ None of the above

2. The graph of a linear function y = mx + 2 goes through the point (4, 0). Which of the following must be true?

   Ⓐ m is negative.
   Ⓑ m = 0
   Ⓒ m is positive
   Ⓓ Cannot be determined.

3. The graph of a linear function y = 2x + b passes through the point (-5, 0). Which of the following must be true?

   Ⓐ b is positive.
   Ⓑ b is negative.
   Ⓒ b = 0
   Ⓓ Cannot be determined.

4. The graph of a linear function y = mx + b goes through the point (0, 0). Which of the following must be true?

   Ⓐ m is positive.
   Ⓑ m is negative.
   Ⓒ m = 0
   Ⓓ b = 0

5. The graph of a linear function y = mx + b includes the points (2, 5) and (9, 5). Which of the following gives the correct values of m and b?

   Ⓐ m = 5 and b = 7
   Ⓑ m = 1 and b = 0
   Ⓒ m = 0 and b = 5
   Ⓓ m = 2 and b = 9

6. **What is the slope of the linear function represented by the (x, y) pairs shown in the table below?**

x	y
0	11
2	7
3	5

Ⓐ $-\frac{1}{2}$

Ⓑ $\frac{1}{2}$

Ⓒ - 2

Ⓓ 2

7. **The graph of a certain linear function includes the points (-4, 1) and (5, 1). Which of the following statements describes the function accurately?**

Ⓐ It is an increasing linear function.
Ⓑ It is a decreasing linear function.
Ⓒ It is a constant function.
Ⓓ It does not form a straight-line relationship.

8. **Which of the following linear functions has the greatest slope?**

Ⓐ

x	y
0	1
2	5
4	9

Ⓑ

x	y
0	3
2	6
4	9

Ⓒ

x	y
0	5
2	7
4	9

Ⓓ

x	y
0	7
2	8
4	9

9. A young child is building a tower of blocks on top of a bench. The bench is 18 inches high, and each block is 3 inches high. Which of the following functions correctly relates the total height of the tower (including the bench) h, in inches, to the number of blocks b?

Ⓐ $h = 3b - 18$
Ⓑ $h = 3b + 18$
Ⓒ $h = 18b - 3$
Ⓓ $h = 18b + 3$

10. Jim owes his parents \$10. Each week, his parents pay him \$5 for doing chores. Assuming that Jim does not earn money from any other source and does not spend any of his money. Which of the following functions correctly relates the total amount of money m, in dollars, that Jim will have to the number of weeks w?

Ⓐ $m = -5w - 10$
Ⓑ $m = -5w + 10$
Ⓒ $m = 5w - 10$
Ⓓ $m = 5w + 10$

11. An amusement park charges \$5 admission and an additional \$2 per ride. Which of the following functions correctly relates the total amount paid p, in dollars, to the number of rides, r?

Ⓐ $p = 2r + 5$
Ⓑ $p = 5r + 2$
Ⓒ $p = 10r$
Ⓓ $p = 7r$

12. Which of the following linear functions has the smallest slope?

Ⓐ

x	y
1	2
4	6
7	8

Ⓑ

x	y
1	3
4	6
7	9

Ⓒ

x	y
1	4
4	6
7	8

Ⓓ

x	y
1	5
4	6
7	7

13. The graph of a linear function includes the points (6, 2) and (9, 4). What is the y intercept of the graph?

Ⓐ (0, 0)
Ⓑ (0, -2)
Ⓒ (3, 0)
Ⓓ (0, 3)

14. Which of the following functions has this set of points as solutions?
{(0, -5), (1, 0), and (4, 15)}

Ⓐ $f(x) = 4x - 15$
Ⓑ $f(x) = 0$
Ⓒ $f(x) = -5$
Ⓓ $f(x) = 5x - 5$

15. A music store is offering a special on CDs. The cost is $20.00 for the first CD and $10.00 for each additional CD purchased.
Which of the following functions represents the total amount in dollars of your purchase where x is the number of CDs purchased?

Ⓐ $f(x) = 10x + 20$
Ⓑ $f(x) = 10(x - 1) + 20$
Ⓒ $f(x) = 20x + 10$
Ⓓ $f(x) = 20(x - 1) + 10$

16. A trainer for a professional football team keeps track of the amount of water players consume throughout practice. The trainer observes that the amount of water consumed is a linear function of the temperature on a given day. The trainer finds that when it is 90°F the players consume about 220 gallons of water, and when it is 76°F the players consume about 178 gallons of water. Fill in the value of m in the function : $y = mx - 50$.

y = _____ x - 50

17. Match each slope and coordinate point with the correct equation

	y=−4x−3	y=7x-5	y=-x+2
(1, 2) slope=7	○	○	○
(-2, 5) slope=-4	○	○	○
(3, -1) slope=-1	○	○	○

18. A cell phone company charges $89.99 for a new phone and then $19.99 per month. What is the slope or rate of change for this situation?

Slope = ________

19. **Plot the (x,y) points shown in the below table on a graph such that it represents a linear function.**

x	y
0	11
2	7
3	5

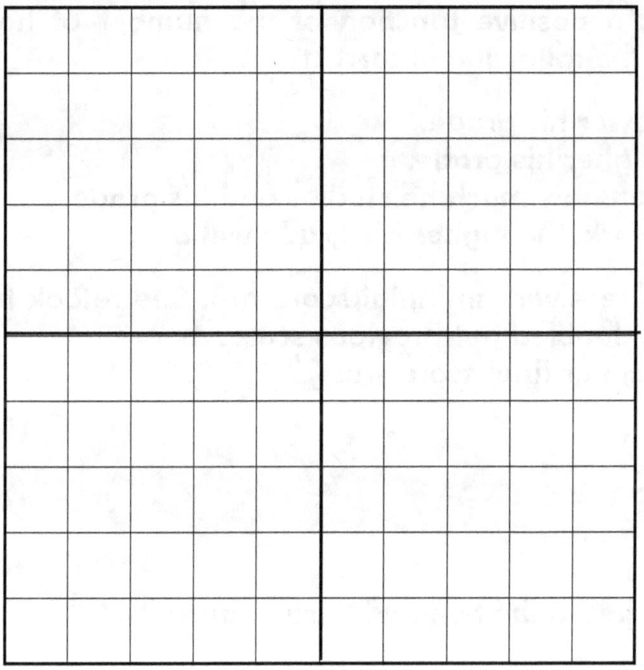

Date of Completion:______________ Score:______________

CHAPTER 3 → Lesson 5: Analyzing Functions

1. Complete the following:
   The cost per copy is a function of the number of copies of any one title purchased. This implies that ______________

   Ⓐ the cost per copy of any one title is always a constant.
   Ⓑ the cost per copy of any one title will change based on the number of copies purchased.
   Ⓒ the cost per copy of any one title is not related to the number of copies purchased.
   Ⓓ None of the above.

2. If a student's math grade is a positive function of the number of hours he spends preparing for a test, which of the following is correct?

   Ⓐ The more he studies, the lower his grade.
   Ⓑ The more he studies, the higher his grade.
   Ⓒ There is no relation between how much he studies and his grade.
   Ⓓ The faster he finishes his work, the higher his grade will be.

3. Mandy took a math quiz and received an initial score of i. She retook the quiz several times and, with each attempt, doubled her previous score.
   After a TOTAL of four attempts, her final score was _______.

   Ⓐ 2i
   Ⓑ 3i
   Ⓒ 2^3i
   Ⓓ None of the above.

4. The graph of a linear function lies in the first and third quadrants. Which of the following must be true?

   Ⓐ It also lies in the second quadrant.
   Ⓑ It also lies in the fourth quadrant.
   Ⓒ It is an increasing function.
   Ⓓ It is a decreasing function.

5. The graph of an increasing linear function crosses the vertical axis at (0, -1). Which of the following CANNOT be true?

   Ⓐ It also lies in the second quadrant.
   Ⓑ It also lies in the fourth quadrant.
   Ⓒ It is an increasing function.
   Ⓓ y intercept is -1.

6. **A linear equation is plotted on the coordinate plane, and its graph is perpendicular to the x-axis. Which of the following best describes the slope of this line?**

Ⓐ **Zero**
Ⓑ **Undefined**
Ⓒ **Negative**
Ⓓ **Positive**

7. **The graph of a decreasing linear function crosses the vertical axis at (0, 3). Which of the following CANNOT be true?**

Ⓐ **The graph lies in the first quadrant.**
Ⓑ **The graph lies in the second quadrant.**
Ⓒ **The graph lies in the third quadrant.**
Ⓓ **The graph lies in the fourth quadrant.**

8. **Which of the following best describes the x & y coordinates of any point in the first quadrant?**

Ⓐ **Both are positive**
Ⓑ **Both are negative**
Ⓒ **One is positive and one is negative**
Ⓓ **None of these**

9. **In the coordinate plane, in which quadrant is the ordered pair, (-3, -6) located?**

Ⓐ **I**
Ⓑ **II**
Ⓒ **III**
Ⓓ **IV**

10. **In the coordinate plane, in which quadrant is the ordered pair, (1, 8) located?**

Ⓐ **I**
Ⓑ **II**
Ⓒ **III**
Ⓓ **IV**

11. In the coordinate plane, in which quadrant is the ordered pair, (6, -9) located?

Ⓐ I
Ⓑ II
Ⓒ III
Ⓓ IV

12. A linear equation is plotted on the coordinate plane, and its graph is parallel to the x-axis. Which of the following best describes the slope of this line?

Ⓐ Zero
Ⓑ Undefined
Ⓒ Negative
Ⓓ Positive

13. Which of the following best describes the x & y coordinates of any point in the third quadrant?

Ⓐ Both are positive
Ⓑ Both are negative
Ⓒ One is positive and one is negative
Ⓓ None of these

14. In a coordinate plane the graphs of two functions are perpendicular. Which of the following best describes the relationship of their slopes?

Ⓐ Their slopes are the same.
Ⓑ Their slopes are both zero.
Ⓒ Their slopes are both one.
Ⓓ One slope is the negative reciprocal of the other.

15. In a coordinate plane the graphs of two linear functions are parallel lines. Which of the following best describes the relationship of their slopes?

Ⓐ Their slopes are the same.
Ⓑ Their slopes are opposites.
Ⓒ The slope of one line is the negative reciprocal of the other.
Ⓓ There is no known relationship between their slopes.

16. **Observe the graph given.**
Match each segment to whether it is increasing or decreasing as per the graph.

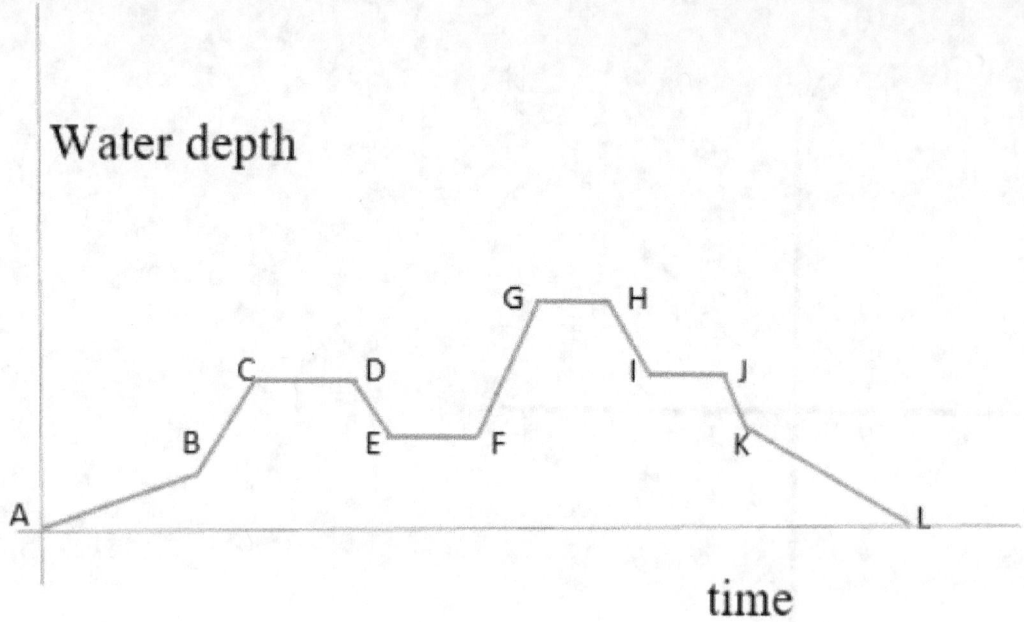

	INCREASING	DECREASING	CONSTANT
A to B	○	○	○
B to C	○	○	○
G to H	○	○	○
I to J	○	○	○
J to K	○	○	○
K to L	○	○	○
C to D	○	○	○
E to F	○	○	○
D to E	○	○	○
F to G	○	○	○
H to I	○	○	○

17. In which interval is this graph increasing?

Write your answer in the box given below

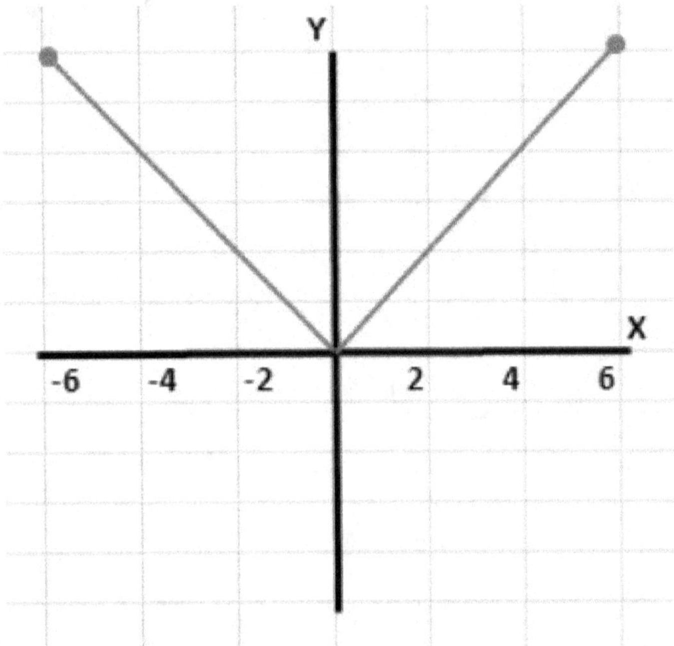

18. **Match each equation with whether or not the function is increasing, decreasing, or constant.**

	INCREASING	DECREASING	CONSTANT
$y=-2x+5$	○	○	○
$y=6$	○	○	○
$y=-\frac{1}{2}x$	○	○	○
$y=5x-9$	○	○	○

End of Functions

ANSWER KEY AND DETAILED EXPLANATION

Chapter 3: Functions

Lesson 1: Functions

<table>
<tr><th>Question No.</th><th>Answer</th><th>Detailed Explanation</th></tr>
<tr><td>1</td><td>D</td><td>In order to be classified as a function each x must map to one and only one value of y.
In the case of the set of ordered pairs {(2, 3), (2, 7), (8, 6)}, the x value of 2 has two different values for y, 3 and 7.
Therefore, this set does not classify as a function.</td></tr>
<tr><td>2</td><td>B</td><td>In the case of
<table><tr><th>x</th><th>y</th></tr><tr><td>1</td><td>7</td></tr><tr><td>3</td><td>8</td></tr><tr><td>4</td><td>7</td></tr></table>for each x there is one and only one value for y.
Therefore, this table represents a function.</td></tr>
<tr><td>3</td><td>A</td><td>If y is a function of x, a particular x value CANNOT be associated with two or more values of y.</td></tr>
<tr><td>4</td><td>A</td><td>In the first graph, for a certain value of x, there are two distinctly different values for y. This is NOT a function.</td></tr>
<tr><td>5</td><td>C</td><td>In the third graph for certain values of x, there are more than one value for y. This is NOT a function.</td></tr>
<tr><td>6</td><td>D</td><td>{(0, 0), (0, 2), (0, 4)} does not qualify as a function because for x = 0, there is more than one value for y.</td></tr>
<tr><td>7</td><td>D</td><td>A line drawn perpendicular to the x-axis will cross the graph exactly once because each x-value corresponds to only one y-value.</td></tr>
</table>

Question No.	Answer	Detailed Explanation
8	D	In a function, for each value of x, there must be one and only one value for y. We already have (2,1) so cannot have another ordered pair where x=2.
9	D	There is already an assigned y value for x=6, x=5, and x=4, but not for x = 3.
10	D	Only this set, [(-7,10), (7,10), (8,9), (9,10)], represents a function because for each value for x, there exists one and only one value for y.
11	A, C	If a set of ordered pairs is a function then it will have each x-value going to only one y-value. If an x-value is paired with more than one y-value, then it is not a function.
12	y=x-3	The rule for this function is y=x-3. To get this you find the slope first. In this case the slope is 1. Then you find out when the y is zero. In this case it is when x is 3. Thus combining them to form the rule.
13		It is a function if each x-value is paired with only one y-value. If y depends linearly on x, in the form y = mx + b, then it is also a function because, in such cases, for each value of x, there is only one value of y.

Lesson 2: Comparing Functions

Question No.	Answer	Detailed Explanation
1	A	Only p = 2(n – 5) meets the requirements to subtract 5 from n and then double the result to get p.
2	B	The slope is 2. The only function with a slope of 2 is y = 2x + 5.
3	B	According to the table, when x increases by 1, y increases by 2. Therefore the slope of the line (m) is 2. The only function offered with a slope of 2 is y = 2x - 4.
4	C	Function A has a slope of $\frac{1}{3}$ and Function B has a slope of $\frac{1}{2}$. Therefore, Function B has a greater slope. To verify, apply the formula: Slope = m $=(y_2-y_1)/(x_2-x_1)$ to each function.
5	C	The slope (m) of Line A $=\frac{9-4}{7-(-1)}=\frac{5}{8}$ The slope (m) of Line B $=\frac{-3-2}{-3-5}=\frac{-5}{-8}=\frac{5}{8}$ Since their slopes are equal, the lines are parallel.
6	A	The slope (m) of Line R $=\frac{4+2}{6+2}=\frac{6}{8}=\frac{3}{4}$ The slope (m) of Line S $=\frac{4-0}{0-3}=\frac{4}{-3}$ Since their slopes are negative reciprocals of each other, the lines are perpendicular.
7	A	y = mx + b $m=\frac{13-1}{-4-2}=\frac{12}{-6}=-2$ y = -2x + b Substitute (x, y) for one point and solve for b. Using (2, 1), we get 1 = -2(2) + b. 1 = -4 + b. 1 + 4 = b. 5 = b; so the y-intercept is (0, 5).
8	C	If you substitute the coordinates of the point (1,9) into each equation, you will find that both equations are satisfied. The graphs of the functions will intersect at the point (1,9) is the correct answer.

Question No.	Answer	Detailed Explanation
9	B	Since their slopes, 3 and $-\frac{1}{3}$, are negative reciprocals of each other, the lines will be perpendicular.
10	A	The slope of line A is 1 and that of line B is 5. Therefore, line B has a steeper slope.
11	B, C	These are showing different representations of a function. In comparing them you need to look at their slopes. Whichever one is higher represents the faster person. In the equations, the slope is found by looking at the number that is paired with the x. From the table or the ordered pairs you have to use the slope formula. (A) CJ's speed = (13-0) / (2-0) = 13/2. Holly's speed = 6. Therefore CJ's speed is greater than Holly's speed, so the conclusion that Holly is faster is wrong. (B) Hank's speed = 7/2. Smith's speed = (5-0) / (4-0) = 5/4. Therefore Hank's speed > Smith's speed. Conclusion is correct. (C) Tyler's speed = (6-0) / (4-0) = 6/4 = 3/2. Omar's speed = 3/2. Therefore Tyler's speed = Omar's speed. Conclusion is correct.
12	Rate of change of function f < [Rate of change of function g]	For the function y = 100x, you know the rate of change is 100, so you need to find out the rate of change of the other one. To do this you will have to find the slope, $\frac{180-0}{3-0}$ which is equal to 60. So in this case we are comparing 60 and 100. Since 60 is smaller than 100 the inequality symbol should be < (less than) for this situation.
13		To determine whether the rate of change is positive or negative you have to look at the slope. The slope is the number paired with the x. If it is a positive number, then the rate of change is positive. If the number is negative, then the rate of change is negative.

Lesson 3: Linear Functions

Question No.	Answer	Detailed Explanation
1	D	The slope of the line through the given points is $\frac{1}{2}$. So slope (m) = $\frac{11 - 7}{k - 6} = \frac{1}{2}$ k-6 = 2(11-7) k-6 = 8 k = 14
2	C	In linear equations, none of the variables is of the second power or higher and they can be put into the form of y = mx + b.
3	B	In the function f(x) = x - 3, the coordinates of (3,0) satisfy the equation and it is written in the form of a linear equation.
4	A	The slope is a constant $\frac{3}{2}$ between any two points and slope is positive; so y increases as x increases. Therefore, it is increasing linearly.
5	D	Only the last statement, the function is nonlinear because it does not have the same slope between different airs of points, is true.
6	B	If it is a function, x has to change as y changes, but if that is true, the line cannot be constant for y unless it is in quadrants I and II, or III and IV, or it is the x-axis.
7	C	If the function is linear, the slope has to be the same between any 2 points. Using the first two points, the slope is $\frac{2}{3}$. Therefore, using the next two points, the slope must also be $\frac{2}{3}$; so $\frac{n - 3}{9 - 3} = \frac{2}{3}$; so 3n-9=18-6 3n=12+9=21 n=7

Question No.	Answer	Detailed Explanation
8	C	$\frac{9-3}{3-0}=\frac{n-9}{9-3}$ $\frac{6}{3}=\frac{n-9}{6}$ $\frac{2}{1}=\frac{n-9}{6}$ n-9=12 n=21
9	D	If the graph lies in quadrants I and II and is non-negative, it cannot reach quadrant IV, although it could also pass through quadrant III.
10	A	Since the slope is 2, y = 2x + 2 is the only function that could fit.
11	B, C	If the graph is a curved line, or the function has a power of x other than 1, then the function is not linear. Therefore, option (A) and (D) are not linear. Option (B) has a power of x equal to 1 only. Therefore, it is linear. Calculate the ratio of change in y to change in x. If the ratio is same for all points, then the graph is a straight line and the function is linear. Option (C) satisfies this condition. Therefore, it is linear.
12	Non Linear	This function will be non-linear because if we rearrange to solve for y, you will still have x to the second power. Thus making it non-linear.
13		If the function has a power of x other than 1, then the function is non-linear. Therefore, first function is non-linear. Second function is a constant function. It is a linear function. Calculate the ratio of change in y to change in x. If the ratio is same for all points, then the graph is a straight line and the function is linear. The table in the third option does not satisfy this condition. Therefore it is a non-linear function. Fourth function has a power of x equal to 1 only. Therefore, it is linear.

Lesson 4: Linear Function Models

Question No.	Answer	Detailed Explanation
1	D	There is not enough information to say any of the choices MUST be true although it may represent a portion of the graph of a constant function.
2	A	$y = mx + 2$ $0 = 4m + 2$ $-2 = 4m$ $-\frac{2}{4} = m$ $-\frac{1}{2} = m$ Therefore, m is negative is true.
3	A	$y = 2x + b$ $0 = 2(-5) + b$ $0 = -10 + b$ $10 = b$ Therefore, b is positive is true.
4	D	All that we know is that the y-intercept is (0,0); i.e. b=0.
5	C	$m = \frac{5 - 5}{9 - 2} = \frac{0}{7} = 0$ $y = mx + b$ $5 = 0 + b$ $5 = b$
6	C	$m = \frac{7 - 5}{2 - 3} = \frac{2}{-1} = -2$
7	C	It is linear and the y value is 1 in the case of the two given points. Therefore, it is a constant function.
8	A	The slopes range between $\frac{1}{2}$ and 2; so 2 is the greatest slope. The first option is the correct answer.
9	B	Height of the bench is 18 in. The height increases by 3 in. for each block added; so $h = 3b + 18$

Question No.	Answer	Detailed Explanation
10	C	He starts out \$10 in debt so b = -10. At any point, he has earned 5w where w is the number of weeks. Therefore, the equation is m = 5w - 10.
11	A	A guest must pay \$5 admission before any rides; so b=5. He pays \$2 per ride so the total paid in rides is 2r. Now add these two amounts and p = 2r + 5
12	D	The slopes range between $\frac{1}{3}$ and $\frac{4}{3}$; so $\frac{1}{3}$ is the smallest slope. The last choice is the correct answer.
13	B	y = mx + b $m = \frac{4-2}{9-6} = \frac{2}{3}$ Substitute (6,2). $2 = 6(\frac{2}{3}) + b$ 2 = 4 + b 2-4 =b -2 = b Therefore, y-intercept is (0, -2)
14	D	The point (0,-5) tells us that b = -5. $m = \frac{15-0}{4-1} = \frac{15}{3} = 5$ Substitute into y = mx + b. f(x) = 5x - 5
15	B	Since you pay \$20 for the first CD, b = 20. If you purchase x CDs and you already paid \$20 for the first one, the remainder would be (x-1) CDs for which you pay \$10 each. Therefore, f(x) = 10(x-1) + 20.
16	3	To find the missing information, we are looking for the rate of change. You will fill this into the slope formula. Using the slope formula: $\frac{220-178}{90-76}$ =3. Thus, the equation will be y=3x-50.

<table>
<tr><th>Question No.</th><th>Answer</th><th>Detailed Explanation</th></tr>
<tr><td>17</td><td></td><td>
<table>
<tr><th></th><th>y=−4x−3</th><th>y=7x-5</th><th>y=-x+2</th></tr>
<tr><td>(1, 2) slope=7</td><td>○</td><td>●</td><td>○</td></tr>
<tr><td>(-2, 5) slope=-4</td><td>●</td><td>○</td><td>○</td></tr>
<tr><td>(3, -1) slope=-1</td><td>○</td><td>○</td><td>●</td></tr>
</table>
Using the slope and an ordered pair, you can find the y-intercept by substituting the x and y values into the equation y = mx + b. Then solve for b to find the equation of the line.

Alternate Explanation: The slope of the line y = mx + b is the coefficient m (the number in front of x). Since the slopes of the three lines are different, you can identify the equations without calculations.</td></tr>
<tr><td>18</td><td>$19.99</td><td>In this case the slope or rate of change is $19.99.</td></tr>
<tr><td>19</td><td>[{"x":0,"y":11}, {"x":2,"y":7}, {"x":3,"y":5}]</td><td>Coordinates are represented as (x,y)

(0,11) means 0 on x axis and 11 on y axis

(2,7) means 2 on x axis and 7 on y axis.

(3,5) means 3 on x axis and 5 on y axis.

Plot the points.</td></tr>
</table>

Lesson 5: Analyzing Functions

Question No.	Answer	Detailed Explanation
1	B	The cost per copy is a function of the number of copies purchased means that the cost per copy changes as the number of copies changes.
2	B	In a positive function, as one variable increases so does the other.
3	C	Her first score was i. She doubled the previous score three times or 2^3i.
4	C	The line could also lie in the second or fourth quadrant, but it definitely has a positive slope and is, consequently, an increasing function.
5	A	The graph cannot lie in the second quadrant.
6	B	m = change in y / change in x. In this case change in x is 0. Division by 0 is undefined.
7	C	The graph cannot lie in the third quadrant.
8	A	In the first quadrant, both x and y are positive.
9	C	If both coordinates are negative, the point lies in Quadrant III.
10	A	If both coordinates are positive, the point lies in Quadrant I.
11	D	If the x coordinate is positive, the point lies on the right side of the y-axis and if the y coordinate is negative, the point lies below the x-axis. This positions the point in Quadrant IV.
12	A	m = change in y / change in x. If line is parallel to x-axis, change in y is 0. If change in y is 0, m = 0 ÷ n = 0, where n is not zero, but is any other real number resulting from the change in x.
13	B	In the third quadrant, both coordinates are negative.
14	D	The slopes of perpendicular lines are negative reciprocals of each other.
15	A	The slopes of parallel lines are the same.

<table>
<tr><th>Question No.</th><th>Answer</th><th>Detailed Explanation</th></tr>
<tr><td>16</td><td></td><td>
<table>
<tr><th></th><th>INCREASING</th><th>DECREASING</th><th>CONSTANT</th></tr>
<tr><td>A to B</td><td>●</td><td></td><td></td></tr>
<tr><td>B to C</td><td>●</td><td></td><td></td></tr>
<tr><td>G to H</td><td></td><td></td><td>●</td></tr>
<tr><td>I to J</td><td></td><td></td><td>●</td></tr>
<tr><td>J to K</td><td></td><td>●</td><td></td></tr>
<tr><td>K to L</td><td></td><td>●</td><td></td></tr>
<tr><td>C to D</td><td></td><td></td><td>●</td></tr>
<tr><td>E to F</td><td></td><td></td><td>●</td></tr>
<tr><td>D to E</td><td></td><td>●</td><td></td></tr>
<tr><td>F to G</td><td>●</td><td></td><td></td></tr>
<tr><td>H to I</td><td></td><td>●</td><td></td></tr>
</table>
If a graph is increasing it will be going from a lower starting point to a higher ending point. If the segment of the graph is decreasing, then its starting point will be higher than the ending point. If the segment of the graph is constant, then the starting point and ending point will be at the same level.</td></tr>
<tr><td>17</td><td>$0 < X < 6$</td><td>Looking at this graph you can see that the graph is increasing from 0 to 6. This means that the starting point of the graph is lower than the ending point. So you would write the interval of $0 < x < 6$.</td></tr>
<tr><td>18</td><td></td><td>
<table>
<tr><th></th><th>INCREASING</th><th>DECREASING</th><th>CONSTANT</th></tr>
<tr><td>$y=-2x+5$</td><td></td><td>●</td><td></td></tr>
<tr><td>$y=6$</td><td></td><td></td><td>●</td></tr>
<tr><td>$y=-\frac{1}{2}x$</td><td></td><td>●</td><td></td></tr>
<tr><td>$y=5x-9$</td><td>●</td><td></td><td></td></tr>
</table>
When looking at just equations, you will have to look at the slope of the line to determine whether it is increasing, decreasing, or constant. The slope is the number that is paired with the x-value. If the slope is negative, then it is decreasing. If the slope is positive , the function is increasing. If there is no slope, then the line is horizontal, making the function a constant.</td></tr>
</table>

Date of Completion:______________ Score:______________

Chapter 4: Geometry

Lesson 1: Transformations of Points and Lines

1. The point (4, 3) is rotated 90° clockwise about the origin. What are the coordinates of the resulting point?

Ⓐ (-3, 4)
Ⓑ (-4, 3)
Ⓒ (4, -3)
Ⓓ (3, -4)

2. A line segment has a length of 9 units. After a certain transformation is applied to the segment, the new segment has a length of 9 units. What was the transformation?

Ⓐ A rotation
Ⓑ A reflection
Ⓒ A translation
Ⓓ Any of the above transformations.

3. The point (2, 4) is rotated 180° clockwise about the origin. What are the coordinates of the resulting point?

Ⓐ (-2, -4)
Ⓑ (-2, 4)
Ⓒ (2, -4)
Ⓓ (2, 4)

4. Two points are located in the (x, y) plane on the opposite sides of the y-axis. After a certain transformation is applied to both points, the two new points end up again on the opposite sides of the y-axis. What was the transformation?

Ⓐ A rotation
Ⓑ A reflection
Ⓒ A translation
Ⓓ A dilation

5. A certain transformation is applied to a line segment. The new segment shifted to the left within the coordinate plane. What was the transformation?

Ⓐ A rotation
Ⓑ A reflection
Ⓒ A translation
Ⓓ It cannot be determined.

6. A line segment with end points (1, 1) and (5, 5) is moved and the new end points are now (1, 5) and (5,1). Which transformation took place?

Ⓐ reflection
Ⓑ rotation
Ⓒ translation
Ⓓ dilation

7. A certain transformation moves a line segment as follows: A (2, 1) moves to A′ (2, -1) and B (5, 3) to B′ (5, -3).
Name this transformation.

Ⓐ Rotation
Ⓑ Translation
Ⓒ Reflection
Ⓓ Dilation

8. After a certain transformation is applied to point (x, y), it moves to (y, -x).
Name the transformation.

Ⓐ Rotation
Ⓑ Translation
Ⓒ Reflection
Ⓓ Dilation

9. A transformation moves the point (0, y) to a new location at (0, -y).
Name this transformation.

Ⓐ Rotation
Ⓑ Translation
Ⓒ Reflection
Ⓓ It could be any one of the three listed.

10. (x,y) is in Quadrant 1. Reflection across the y-axis would move it to a point with the following coordinates.

Ⓐ (x, y)
Ⓑ (-x, -y)
Ⓒ (x, -y)
Ⓓ (-x, y)

11. **Mark TRUE or FALSE based on the description of the transformation.**

	TRUE	FALSE
If you graph a point A (3,2). The point gets translated 10 units down, it will end up at A'(3,12)	○	○
Point A (-2,4) is reflected over the y-axis. The new ordered pair will be A'(2,4).	○	○
Line AB is 3 units long. After it is rotated 90°counter clockwise, the line segment will now be 3 units long.	○	○
Point A(5,8) is translated 3 units to the right. It is now located at A'(8,8)	○	○

12. **Enter the correct operation that will describe the rule for the translation left 3 units and up 4 units?**

(x,y) --> (x ☐ 3, y ☐ 4)

13. **Circle the graph that represents a rotation.**

Ⓐ
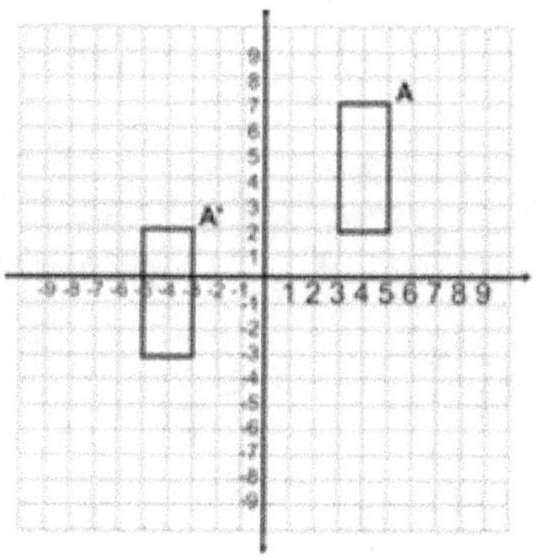

Ⓑ
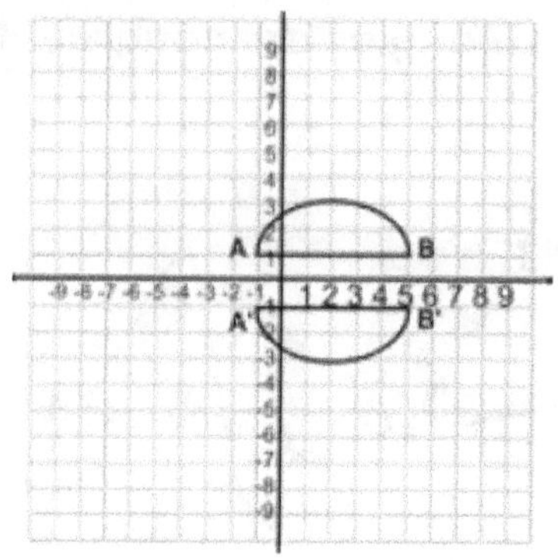

Ⓒ
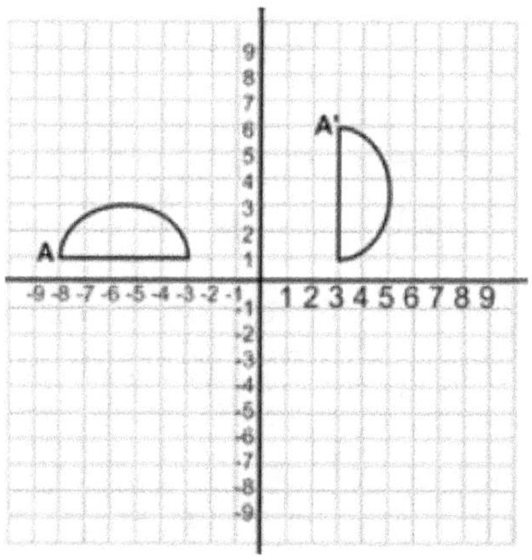

CHAPTER 4 → Lesson 2: Transformations of Angles

1. $\triangle ABC$ is reflected across the x-axis.
Which two angles are congruent?

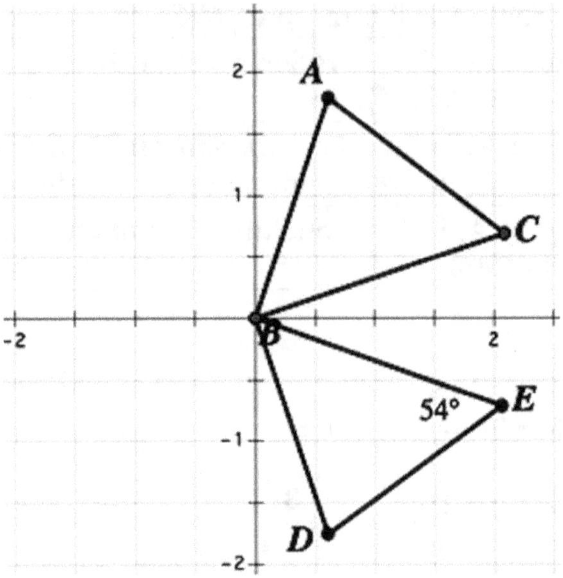

Ⓐ ∠A and ∠C
Ⓑ ∠A and ∠E
Ⓒ ∠C and ∠D
Ⓓ ∠C and ∠E

2. $\triangle ABC$ is rotated 90°.
Which two angles are congruent?

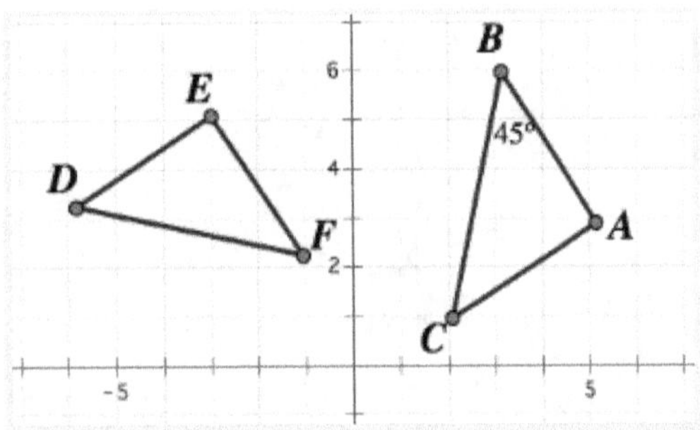

Ⓐ ∠A and ∠C
Ⓑ ∠B and ∠E
Ⓒ ∠C and ∠D
Ⓓ ∠C and ∠F

3. **What rigid transformation should be used to prove $\angle ABC \cong \angle XYZ$?**

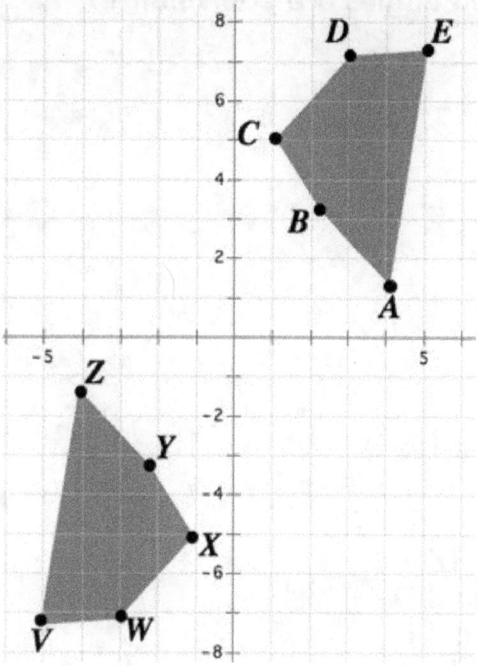

Ⓐ **Reflection**
Ⓑ **Rotation**
Ⓒ **Translation**
Ⓓ **None of the above**

4. **What rigid transformation should be used to prove $\angle ABC \cong \angle XYZ$?**

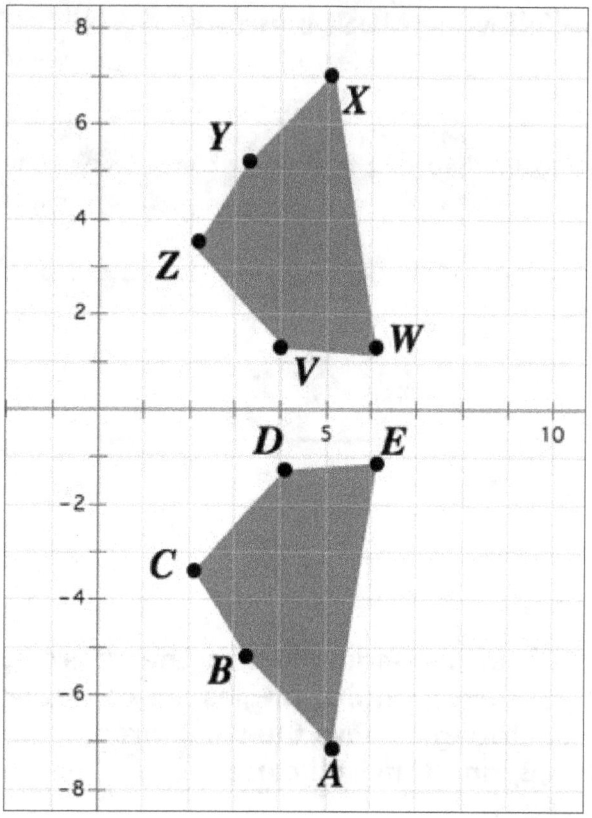

Ⓐ **Reflection**
Ⓑ **Rotation**
Ⓒ **Translation**
Ⓓ **None of the above**

5. **ΔABC is rotated 90°.**
Which two angles are equivalent?

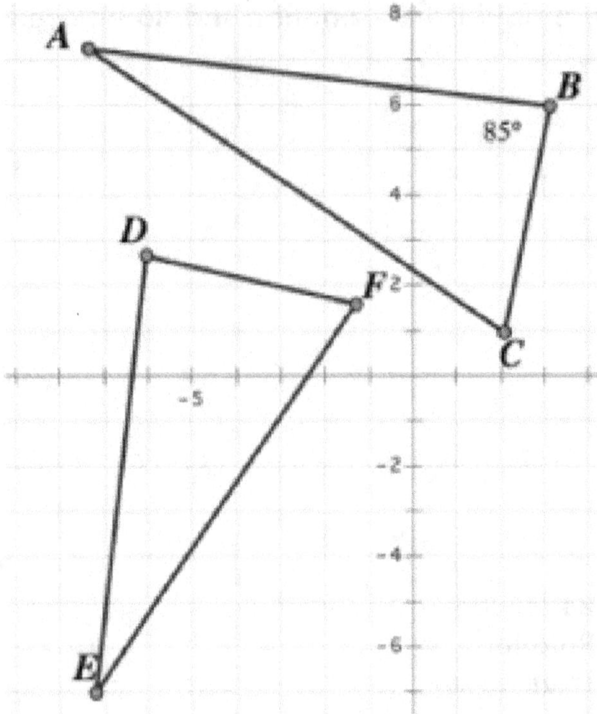

Ⓐ ∠ A and∠ F
Ⓑ ∠ B and∠ D
Ⓒ ∠ B and∠ F
Ⓓ ∠ C and∠ D

6. **If all the triangles below are the result of one or more rigid transformations, which of the following MUST be true?**

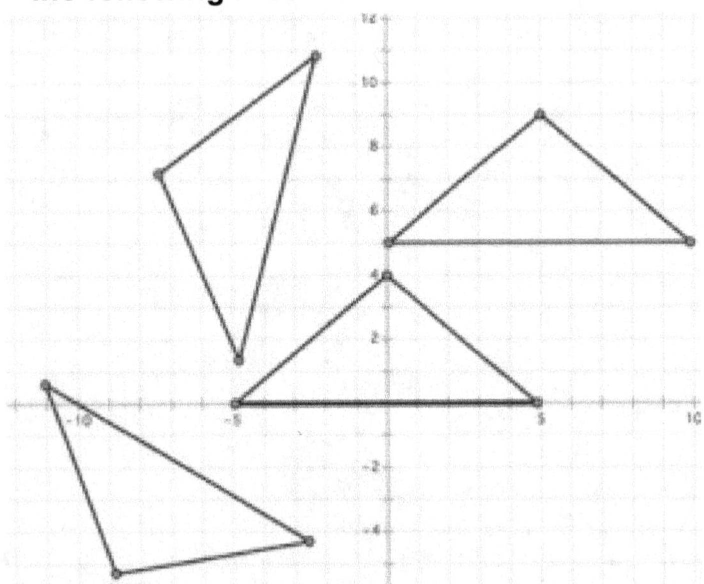

Ⓐ **All corresponding line segments are congruent.**
Ⓑ **All corresponding angles are congruent.**
Ⓒ **All triangles have the same area.**
Ⓓ **A,B, and C are all correct.**

7. **What rigid transformation should be used to prove $\angle ABC \cong \angle DEF$?**

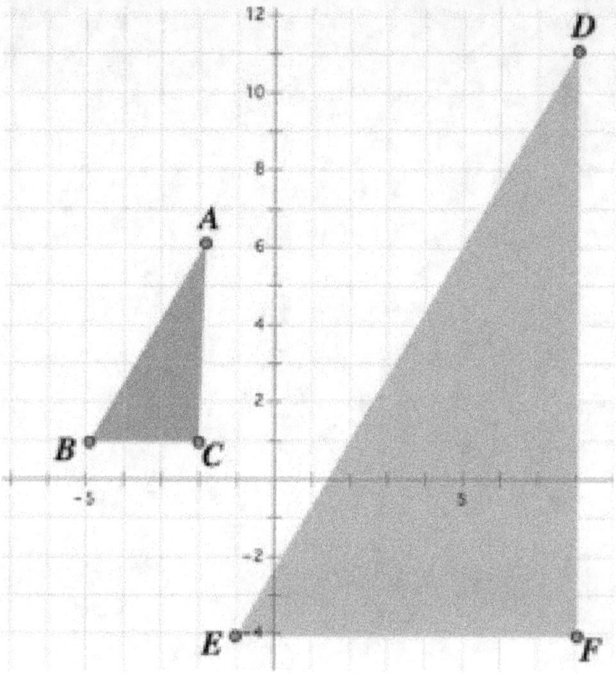

Ⓐ **Reflection**
Ⓑ **Rotation**
Ⓒ **Translation**
Ⓓ **None of the above**

8. **A company is looking to design a new logo, which consists only of transformations of the angle below:**

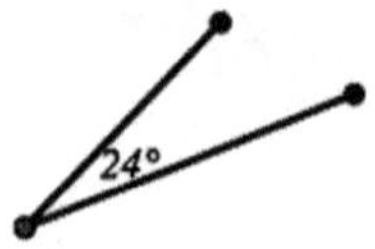

Which logo meets the company's demand?

Ⓐ 

Ⓑ 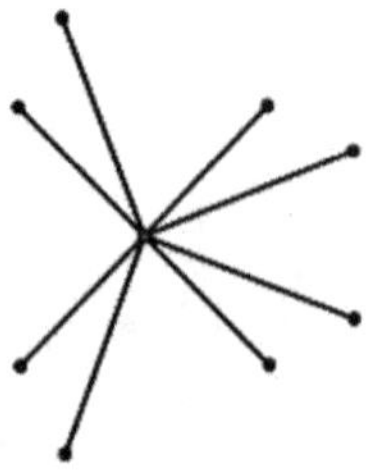

Ⓒ 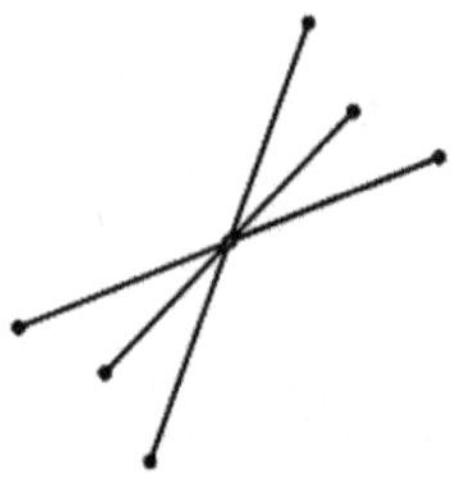

Ⓓ **All of the above**

9. **The angle ∠ AOB is 45° and has been rotated 120° around point C. What is the measure of the new angle ∠ XYZ?**

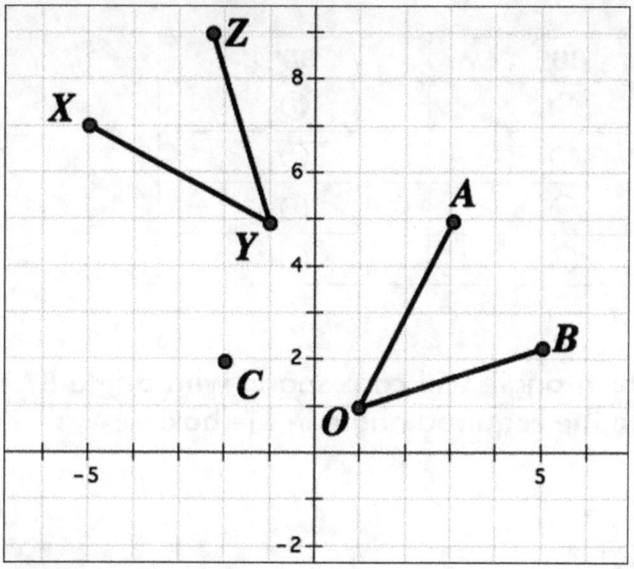

Ⓐ **30°**
Ⓑ **45°**
Ⓒ **90°**
Ⓓ **120°**

10. **Find the measure of ∠ ABC.**

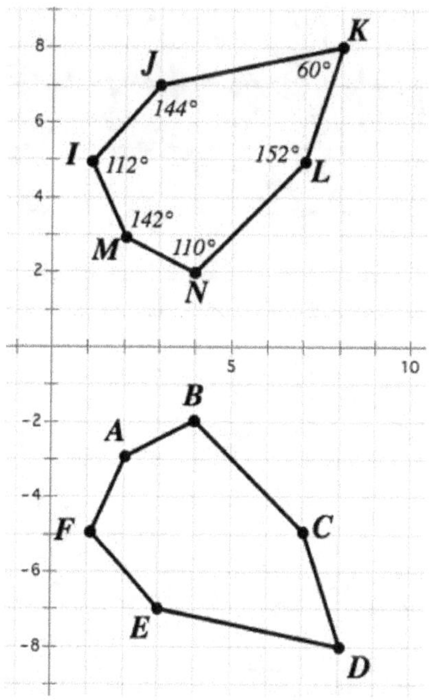

Ⓐ **110°**
Ⓑ **112°**
Ⓒ **142°**
Ⓓ **144°**

11. **Select the angle measure that corresponds with each transformation. Your preimage has an angle measure of 30°.**

	30°	60°	90°
Translation	○	○	○
Reflection	○	○	○
Rotation	○	○	○
Dilation	○	○	○

12. **In the figure below $\Delta ABC \cong \Delta DEF$. . Which angle will correspond with angle B? Type the letter (in capitals) which corresponds to the required angle in the box.**

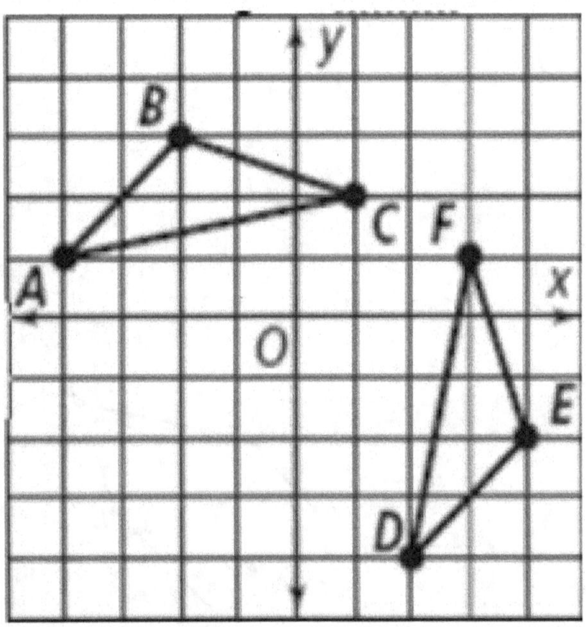

13. **Triangle ABC is rotated 90° counterclockwise. Which two angles would be congruent?**

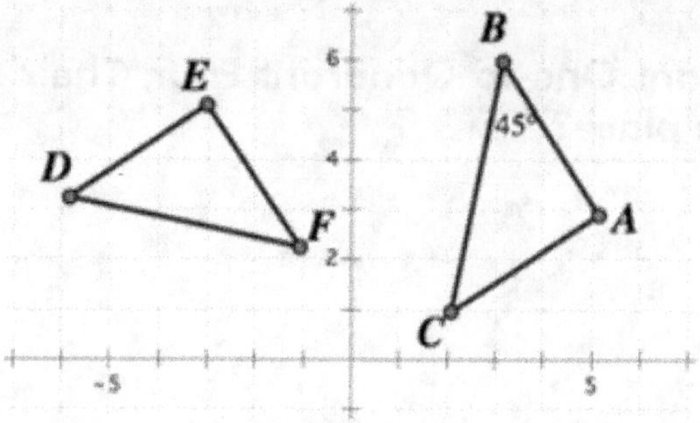

Ⓐ **Angle A and Angle C**
Ⓑ **Angle B and Angle E**
Ⓒ **Angle C and Angle D**
Ⓓ **Angle C and Angle F**

CHAPTER 4 → Lesson 3: Transformations of Parallel Lines

1. Two parallel line segments move from Quadrant One to Quadrant Four. Their slopes do not change. What transformation has taken place?

Ⓐ Reflection
Ⓑ Translation
Ⓒ Dilation
Ⓓ This is not a transformation.

2. Two parallel line segments move from Quadrant One to Quadrant Four. Their slopes change from a positive slope to a negative slope. What transformation has taken place?

Ⓐ Reflection
Ⓑ Rotation
Ⓒ Translation
Ⓓ It could be either a rotation or a reflection.

3. Two parallel line segments move from Quadrant One to Quadrant Two. Their slopes change from a negative slope to a positive slope. What transformation has taken place?

Ⓐ Reflection
Ⓑ Rotation
Ⓒ Translation
Ⓓ It could be either a rotation or a reflection.

4. Line L is translated along segment $\overline{AB}$ to create line L' . Will L and L' ever intersect?

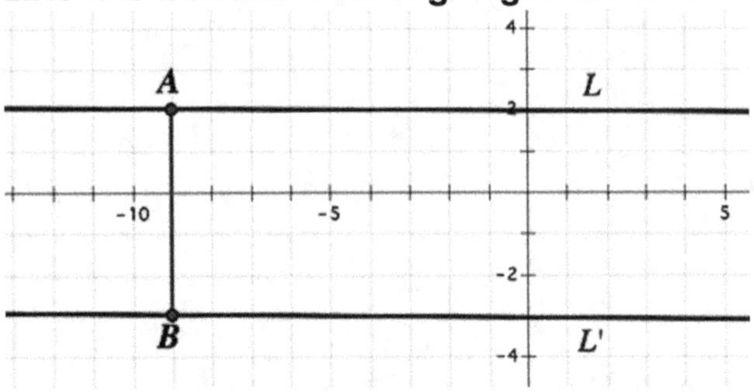

Ⓐ Yes, line L' is now the same as L.
Ⓑ Yes, parallel lines always eventually intersect.
Ⓒ No, every point on L' will always have a corresponding point the distance of $\overline{AB}$ away from L'.
Ⓓ No, the translation along $\overline{AB}$ does not change the slope from L to L', and lines with the same slope never intersect.

5. Line *L* is translated along segment $\overline{AB}$ to create line *L'*. Will *L* and *L'* ever intersect?

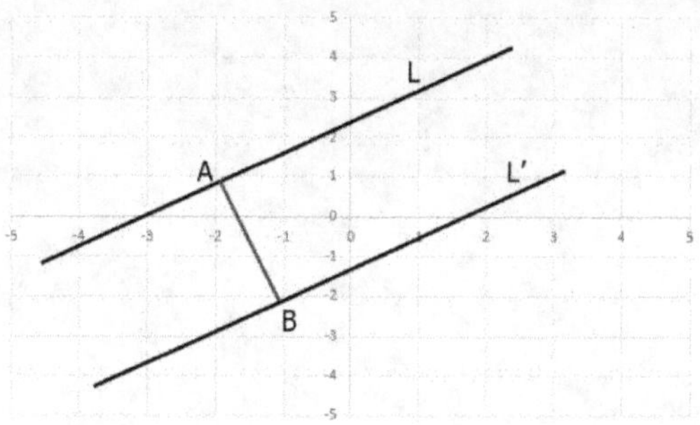

Ⓐ Yes, line *L'* is now the same as *L*.
Ⓑ Yes, parallel lines always eventually intersect.
Ⓒ Yes, every point on *L'* will not always have a corresponding point the distance of AB away from *L'*.
Ⓓ No, the translation along the line segment *AB* does not change the slope from *L* to *L'*, and lines with the same slope never intersect.

6. Line *L* is translated along ray *AC* to create line *L'*. What do you know about the relationship between line *L* and *L'*?

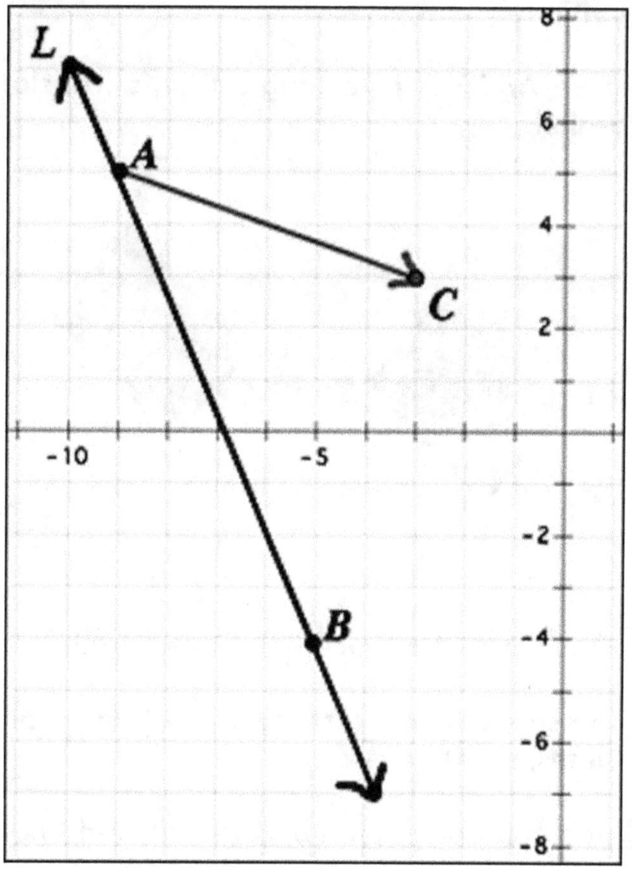

Ⓐ The lines intersect at least once.
Ⓑ The lines are exactly the same.
Ⓒ The lines are parallel.
Ⓓ None of the above.

7. How many lines can be drawn through point C that are parallel to line L?

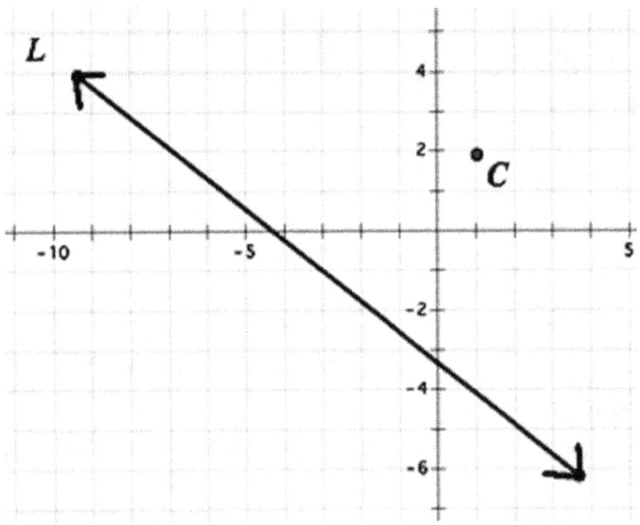

Ⓐ None
Ⓑ One
Ⓒ Two
Ⓓ Infinitely many

8. Figure ABCD is a rectangle that is translated to create $WXYZ$. Prove $\overline{XY}$ and $\overline{WZ}$ are parallel.

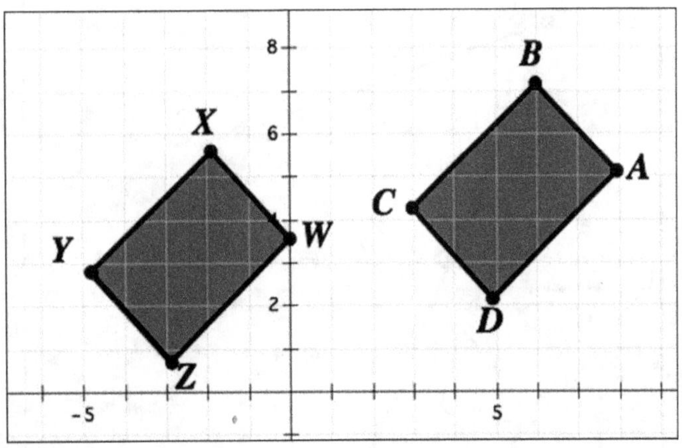

Ⓐ The sides of all rectangles are parallel.

Ⓑ Since $ABCD$ is a rectangle, $\overline{AD} \parallel \overline{BC}$. Translations map parallel lines to parallel lines, so $\overline{XY} \parallel \overline{WZ}$.

Ⓒ It cannot be proven because $\overline{XY}$ and $\overline{WZ}$ are perpendicular.

Ⓓ It cannot be proven because the angle of rotation is not given.

9. **The two parallel lines shown are rotated 180° about the origin. What is the result of this transformation?**

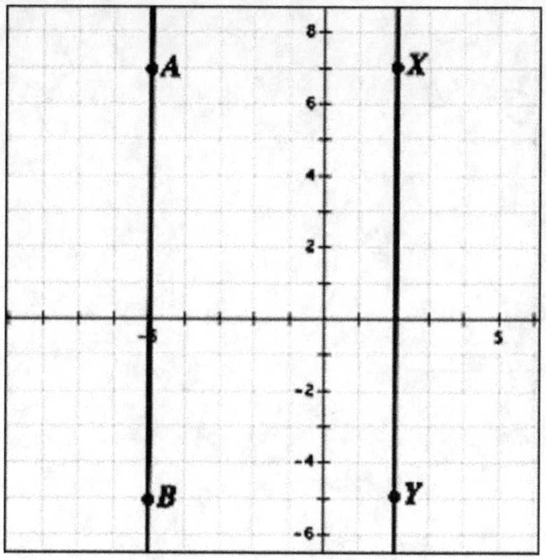

Ⓐ

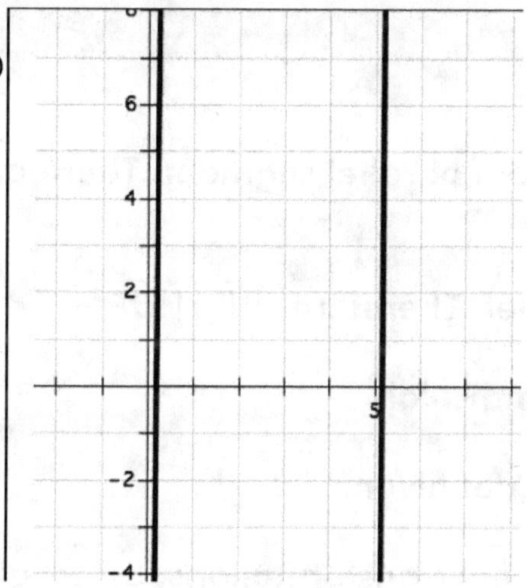

Ⓑ

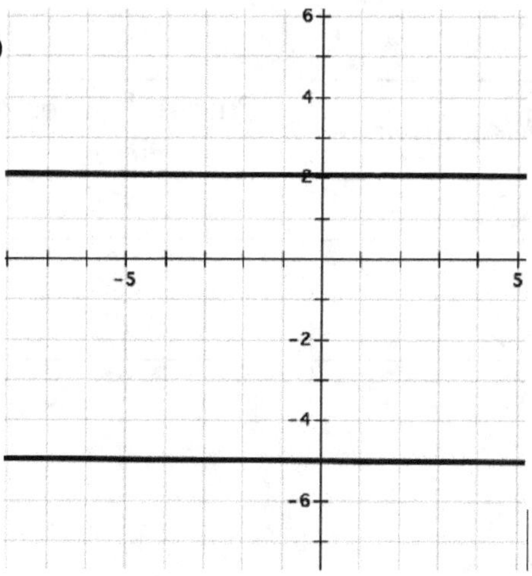

Ⓒ

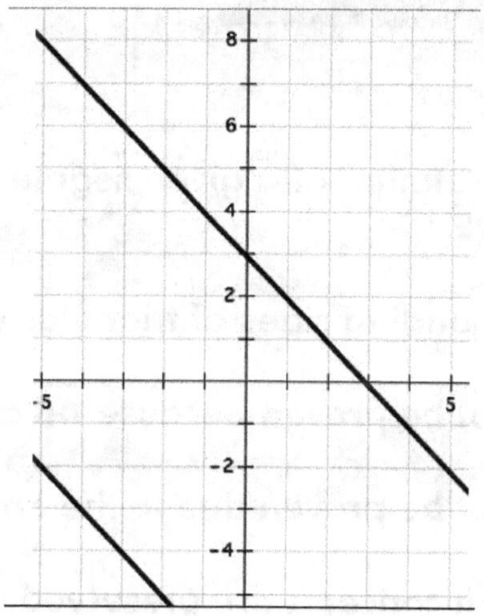

Ⓓ

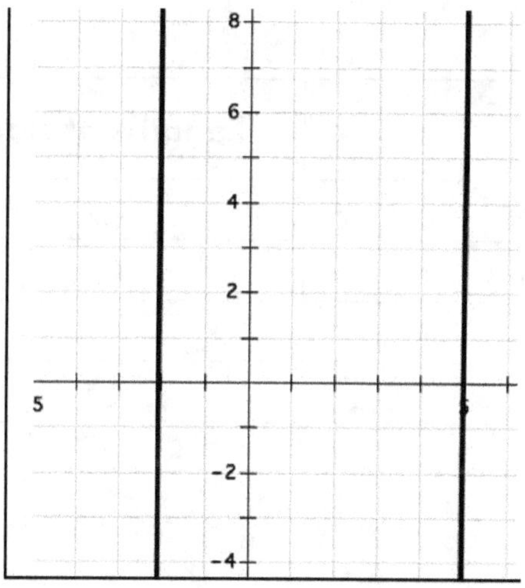

10. Figure *DEF* is the result of 180° rotation around the origin of figure *ABC*. Prove $\overline{AB}$ and $\overline{DE}$ are parallel.

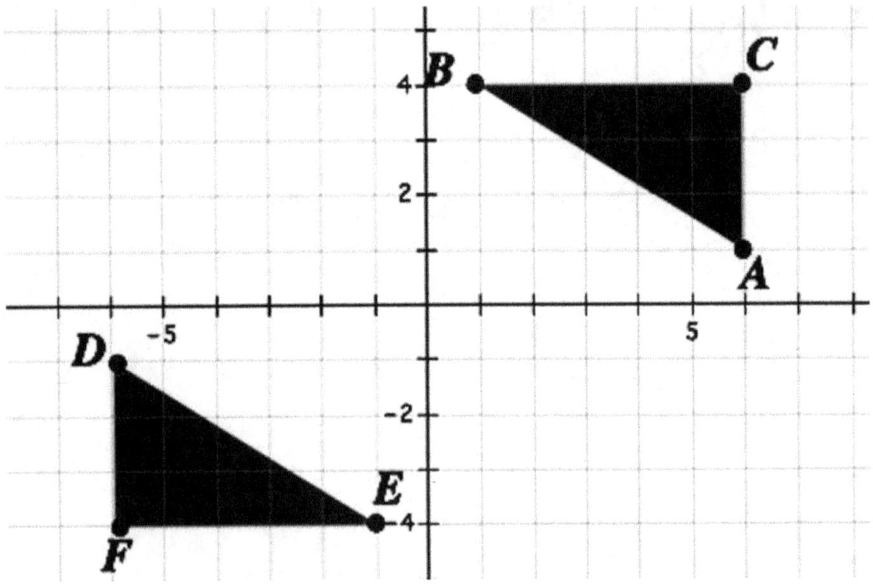

Ⓐ A 180° rotation of a given segment always maps to a parallel segment. Therefore $\overline{AB} \parallel \overline{DE}$.

Ⓑ Corresponding sides of triangles are always parallel. Therefore, $\overline{AB} \parallel \overline{DE}$.

Ⓒ It cannot be proven because $\overline{AB}$ and $\overline{DE}$ are perpendicular.

Ⓓ It cannot be proven due to the angle of rotation is not given.

11. Select what concepts are preserved under these different transformations. Select all that apply.

	Lengths of sides	Angle Measures	Parallel Sides on Figure
Translation	○	○	○
Reflection	○	○	○
Rotation	○	○	○
Dilation	○	○	○

12. Figure ABCD undergoes the shown transformation. The slope of $\overline{AC}$ is $-\frac{1}{3}$. What is the slope of A'C'?

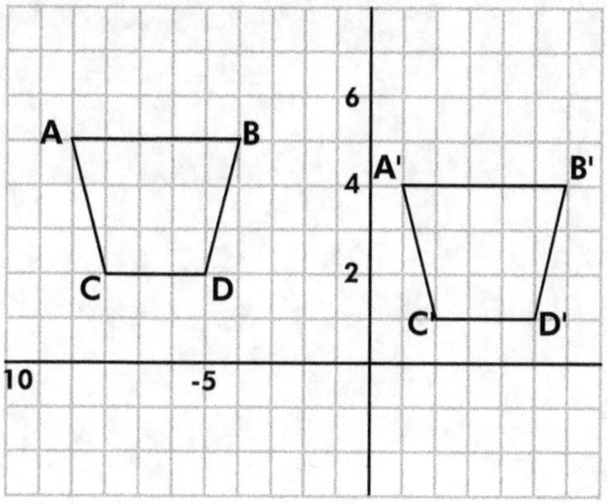

Slope of A'C' is ☐

13. Select the one that correctly shows the parallel lines that have been correctly reflected over the y-axis.

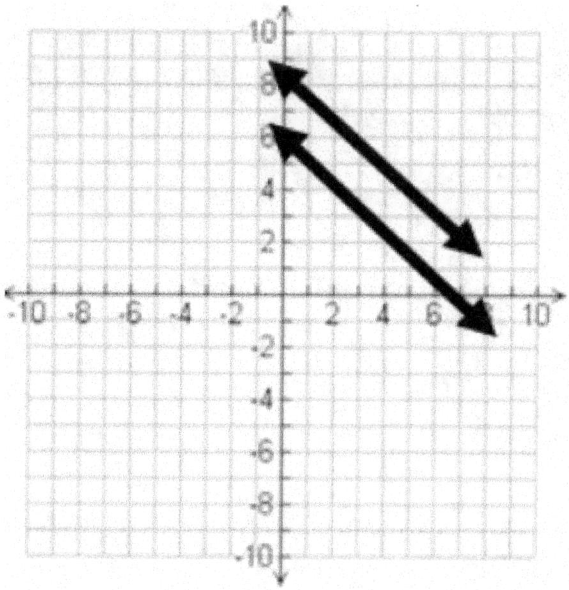

Ⓐ

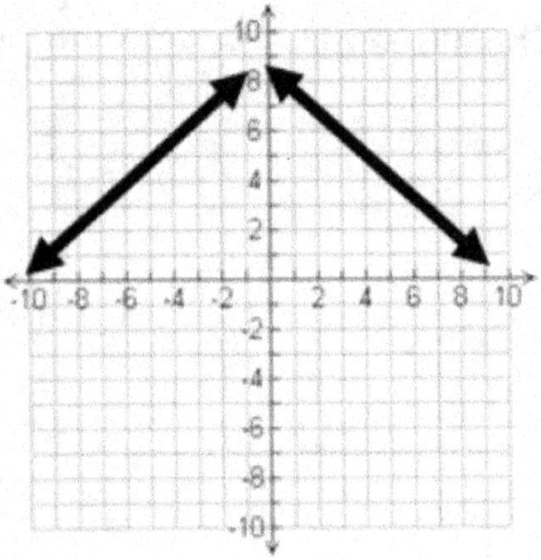

Ⓑ

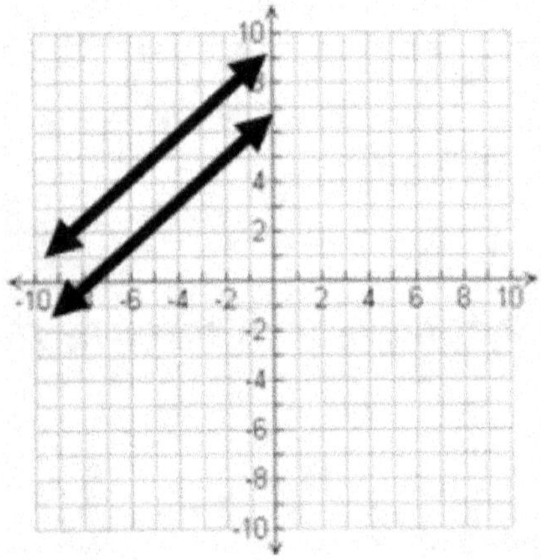

Ⓒ

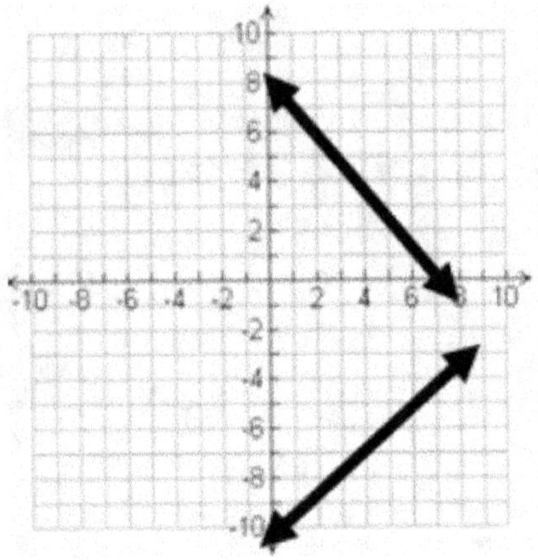

CHAPTER 4 → Lesson 4: Transformations of Congruency

1. Which of the following examples best represents congruency in nature?

Ⓐ A mother bear and her cub.
Ⓑ The wings of a butterfly.
Ⓒ The tomatoes were picked from my garden.
Ⓓ The clouds in the sky.

2. If triangle ABC is drawn on a coordinate plane and then reflected over the vertical axis, which of the following statements is true?

Ⓐ The reflected triangle will be similar ONLY to the original.
Ⓑ The reflected triangle will be congruent to the original.
Ⓒ The reflected triangle will be larger than the original.
Ⓓ The reflected triangle will be smaller than the original.

3. What transformation was applied to the object in quadrant 2 to produce the image shown in the graph below?

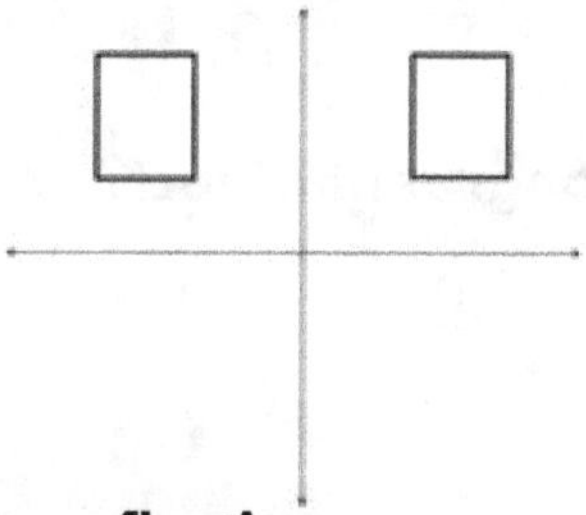

Ⓐ reflection
Ⓑ rotation
Ⓒ translation
Ⓓ not enough information

4. What transformation was applied to the object in quadrant 2 to produce the image shown in the graph below?

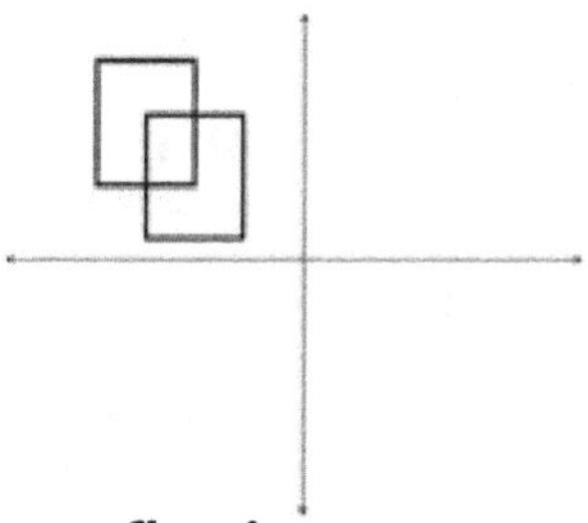

Ⓐ reflection
Ⓑ rotation
Ⓒ translation
Ⓓ not enough information

5. What is NOT true about the graph below?

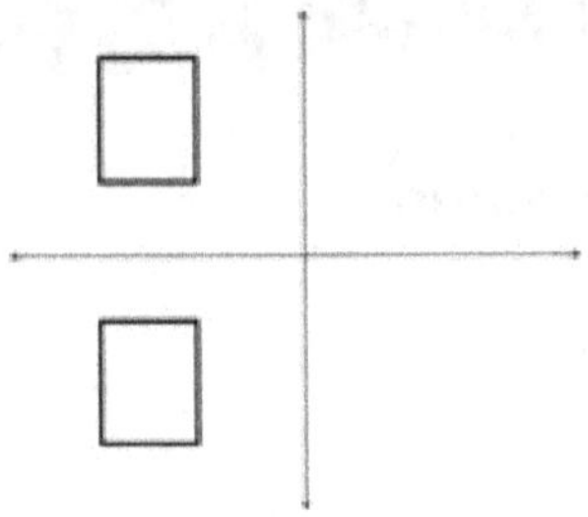

Ⓐ The object in quadrant 3 could not be a reflection of the object in quadrant 2.
Ⓑ The object in quadrant 3 could be a translation of the object in quadrant 2.
Ⓒ The two objects are congruent.
Ⓓ The object in quadrant 3 could not be a dilation of the object in quadrant 2.

6. Finish the statement. Two congruent objects ________________.

Ⓐ have the same dimensions.
Ⓑ have different measured angles.
Ⓒ are not the same shape.
Ⓓ only apply to two-dimensional objects.

7. What transformations can be applied to an object to create a congruent object?

Ⓐ all transformations
Ⓑ dilation and rotation
Ⓒ translation and dilation
Ⓓ reflection, translation, and rotation

8. A figure formed by rotation followed by reflection of an original triangle will be ____.

Ⓐ similar only and not congruent to the original.
Ⓑ congruent to the original.
Ⓒ smaller than the original.
Ⓓ larger than the original.

9. Which of the following is NOT a characteristic of congruent triangles?

Ⓐ They have three pairs of congruent sides.
Ⓑ They have three pairs of congruent angles.
Ⓒ Their areas are equal.
Ⓓ They have four pairs of proportional sides

10. Which of the following letters looks the same after a reflection followed by a 180° rotation.

Ⓐ P
Ⓑ O
Ⓒ F
Ⓓ None of the above.

11. Select all that apply to this transformation.

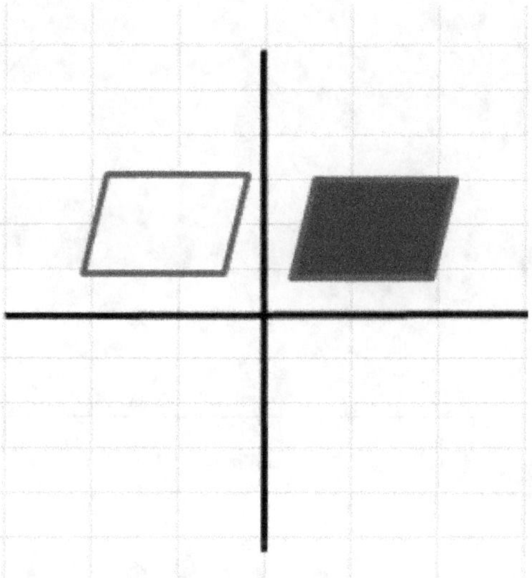

	Apply	Doesn't Apply
the two shapes are congruent	○	○
the two shapes are not congruent	○	○
the two shapes are similar	○	○
the two shapes have the same size	○	○
one shape is rotated from the other shape	○	○
one shape is reflected from the other shape	○	○
one shape is translated from the other shape	○	○

12. **Quadrilateral ABCD is translated 5 units to the left and 4 units down. Which congruent quadrilateral matches this transformation?**

Write your answer in the box given below

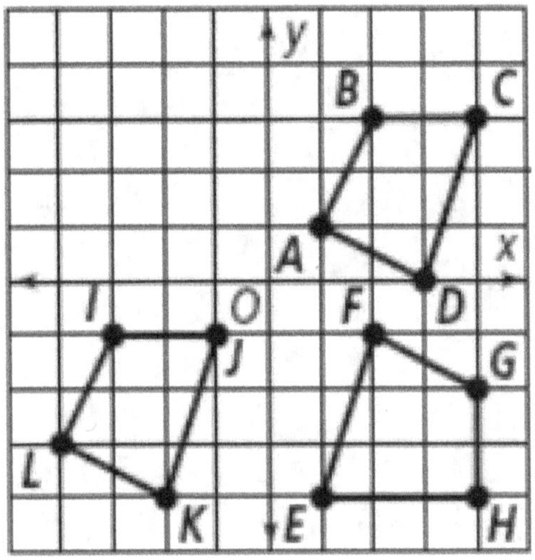

Date of Completion:____________ Score:____________

CHAPTER 4 → Lesson 5: Analyzing Transformations

1. Which of the following transformations could transform triangle A to triangle B?

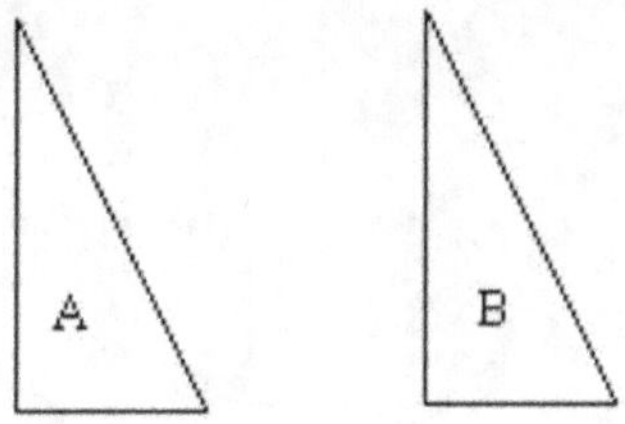

Ⓐ Rotation
Ⓑ Reflection
Ⓒ Translation
Ⓓ Dilation

2. Which of the following transformations could transform triangle A to triangle B?

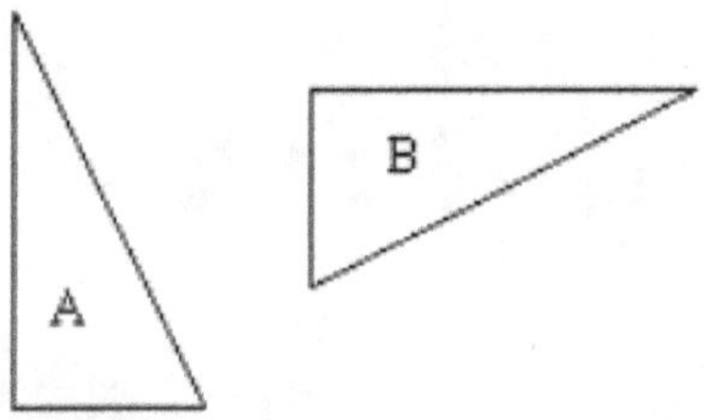

Ⓐ Rotation
Ⓑ Reflection
Ⓒ Translation
Ⓓ Dilation

3. Which of the following transformations could transform triangle A to triangle B?

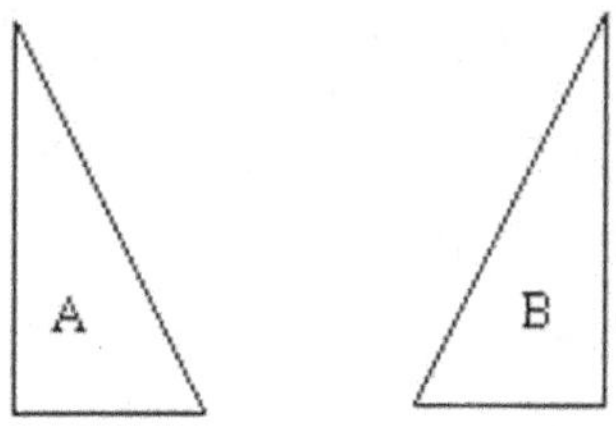

Ⓐ Rotation
Ⓑ Reflection
Ⓒ Translation
Ⓓ Dilation

4. **Which of the following transformations could transform triangle A to triangle B in a single step?**

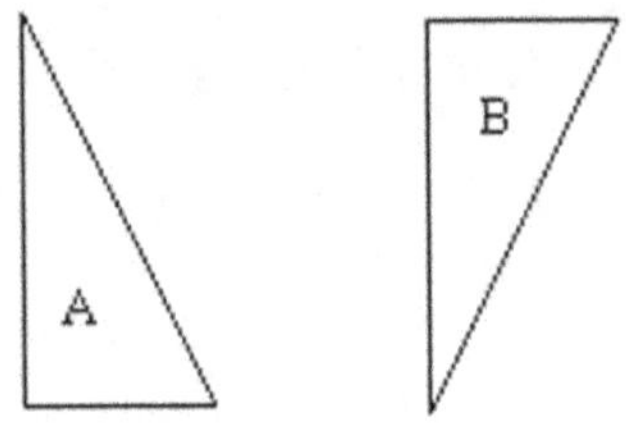

Ⓐ **Rotation**
Ⓑ **Reflection**
Ⓒ **Translation**
Ⓓ **None of the above**

5. **Which of the following transformations could transform triangle A to triangle B?**

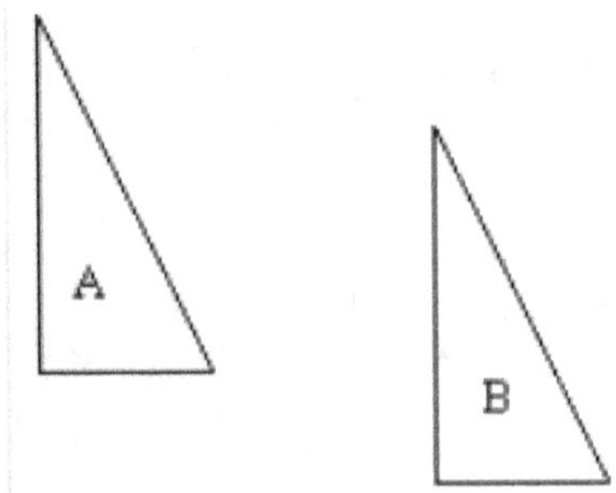

Ⓐ **Rotation**
Ⓑ **Reflection**
Ⓒ **Translation**
Ⓓ **Dilation**

6. **Which of the following transformations could transform triangle A to triangle B?**

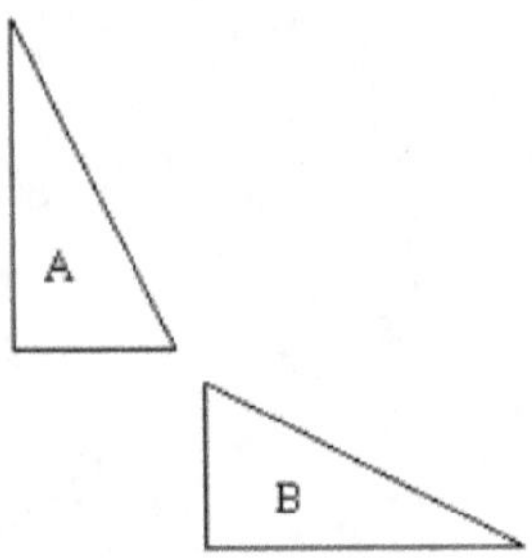

Ⓐ **Rotation**
Ⓑ **Reflection**
Ⓒ **Translation**
Ⓓ **None of the above**

7. Which of the following sequences of transformations could transform triangle A to triangle B?

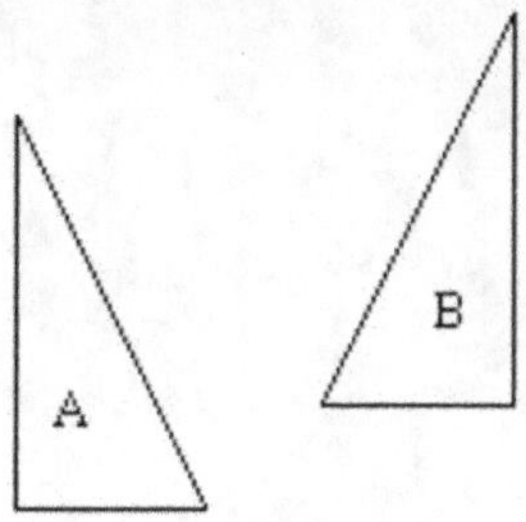

Ⓐ A reflection followed by a translation
Ⓑ A reflection followed by another reflection
Ⓒ A rotation followed by a translation
Ⓓ A dilation followed by a translation

8. Which of the following transformations does NOT preserve congruency?

Ⓐ Rotation
Ⓑ Translation
Ⓒ Reflection
Ⓓ Dilation

9. Consider the triangle with vertices (1, 0), (2, 5) and (-1, 5). Find the vertices of the new triangle after a reflection over the vertical axis followed by a reflection over the horizontal axis.

Ⓐ (-1, 0), (1, 5) and (-2, 5)
Ⓑ (-1, 0), (-2, -5) and (1, -5)
Ⓒ (1, 0), (-2, 5) and (1, 5)
Ⓓ (1, 0), (-2, -5) and (1, -5)

10. Translate the triangle with vertices (1, 0), (2, 5), and (-1, 5), 3 units to the left. Which of the following ordered pairs represent the vertices of the new triangle?

Ⓐ (-2, 0), (-4, 5) and (-1, 5)
Ⓑ (4, 0), (5, 5) and (2, 5)
Ⓒ (-1, 2), (2, 2) and (1, -3)
Ⓓ (-1, 8), (2, 8) and (1, 3)

11. Select the coordinates that will correspond with each transformation for point A.

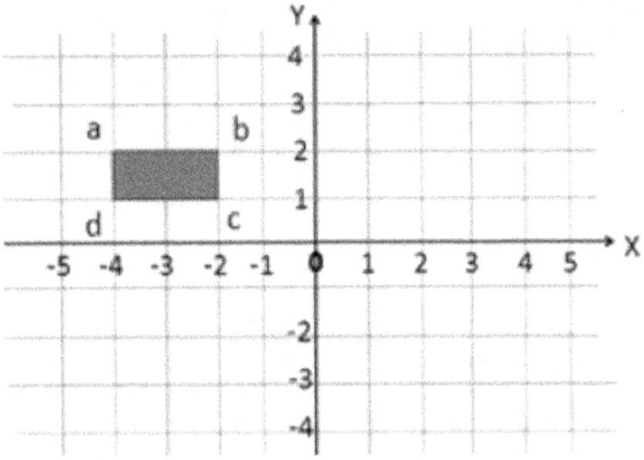

	A(4,-2)	A(-2,1)	A(-4,-2)
Translation (x+2,y-1)	○	○	○
Rotation 180°	○	○	○
Reflection over x-axis	○	○	○

12. In the coordinate plane shown,ΔABC has vertices A(7, 6), B(4, 2), and C(10, 2). What scale factor was used on ΔABC to get ΔDEF.

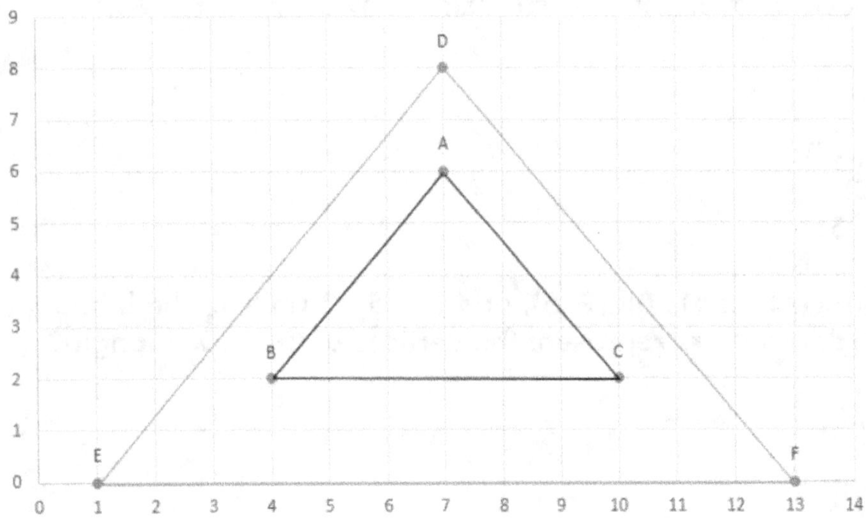

CHAPTER 4 → Lesson 6: Transformations of Similarity

1. **Which of the following transformations could transform triangle A to triangle B?**

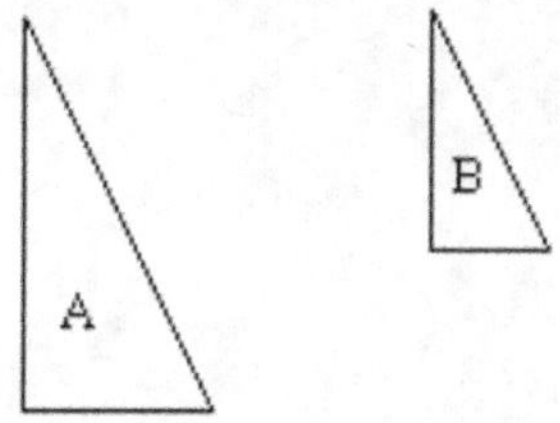

Ⓐ **Rotation**
Ⓑ **Reflection**
Ⓒ **Translation**
Ⓓ **Dilation**

2. **What transformations have been applied to the large object to render the results in the graph below?**

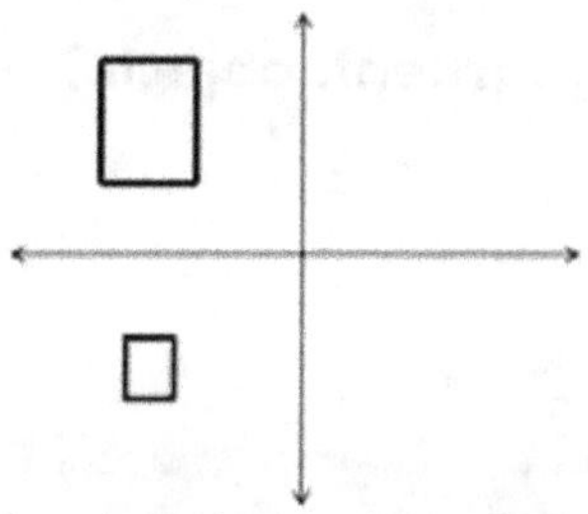

Ⓐ **rotation and dilation**
Ⓑ **rotation and translation**
Ⓒ **translation and dilation**
Ⓓ **None of the above**

3. **What transformations have been applied to the large object to render the results in the graph below?**

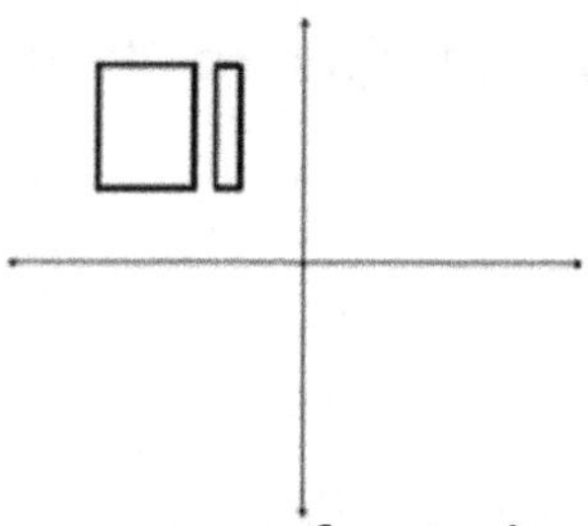

Ⓐ **no transformation**
Ⓑ **reflection and dilation**
Ⓒ **translation and dilation**
Ⓓ **rotation and dilation**

4. Which graph represents reflection over an axis and dilation?

Ⓐ 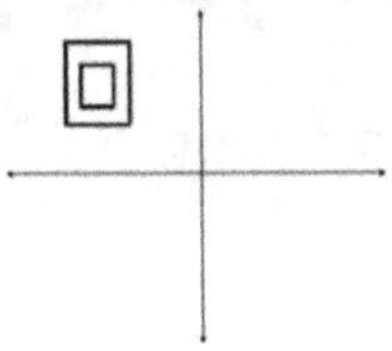

Ⓑ 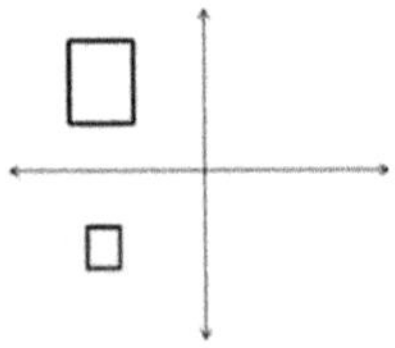

Ⓒ 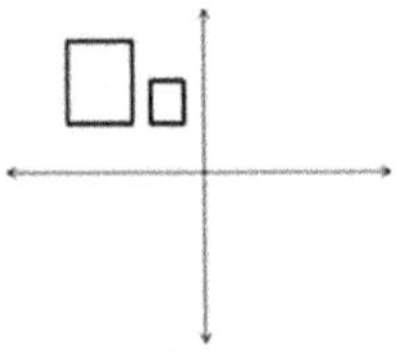

Ⓓ None of the above.

5. What transformation is necessary to have two similar, but not congruent, objects?

Ⓐ Rotation
Ⓑ Translation
Ⓒ Dilation
Ⓓ Reflection

6. Finish this statement. Two similar objects _______________.

Ⓐ have proportional dimensions.
Ⓑ are always congruent.
Ⓒ have different measured angles.
Ⓓ can be different shapes.

7. If a point, P, on a coordinate plane moves from (9, 3) to P′ (3, -9) and then to P″ (0, -9), what transformations have been applied?

Ⓐ Dilation followed by Translation
Ⓑ Translation followed by Dilation
Ⓒ Rotation followed by Translation
Ⓓ Translation followed by Rotation

8. Consider Triangle ABC, where AB = 5, BC = 3, and AC = 6, and Triangle WXY, where WX = 10, XY = 6, and WY = 12.
Assume ABC is similar to WXY.
Which of the following represents the ratio of similarity?

Ⓐ 1 : 2
Ⓑ 5 : 6
Ⓒ 3 : 10
Ⓓ 6 : 20

9. Rectangle A is 1 unit by 2 units.
Rectangle B is 2 units by 3 units.
Rectangle C is 2 units by 4 units.
Rectangle D is 3 units by 6 units.
Which rectangle is not similar to the other three rectangles?

Ⓐ A
Ⓑ B
Ⓒ C
Ⓓ D

10. If triangle ABC is similar to triangle WXY and AB = 9, BC = 7, AC = 14, WX = 27, and XY = 21.
Find WY.

Ⓐ 44
Ⓑ 43
Ⓒ 42
Ⓓ 41

11. Select the transformations used to map the pre-image onto the image. Also, indicate whether the transformation preserves congruence or only similarity.

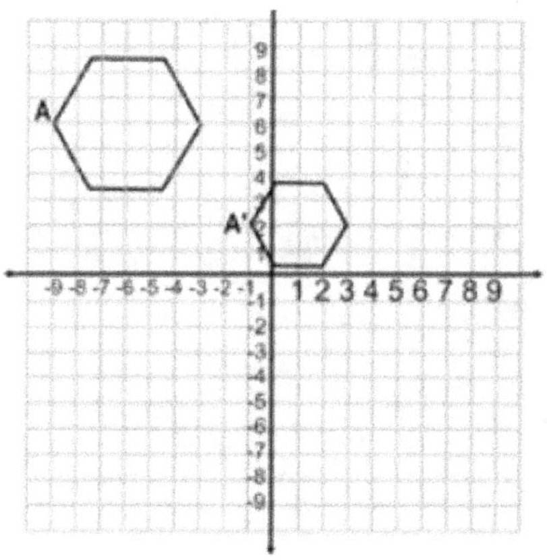

	Used	Similar only	Congruent
Translation	○	○	○
Rotation	○	○	○
Reflection	○	○	○
Dilation	○	○	○

12. The 2 figures are similar. What is the height of the 2nd figure? Write your answer in the box below.

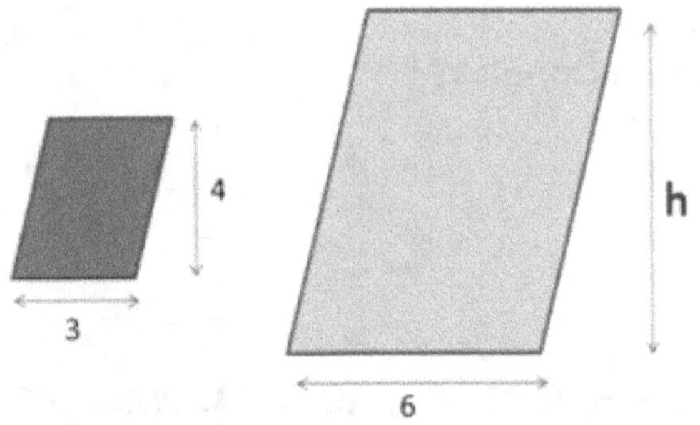

CHAPTER 4 → Lesson 7: Interior & Exterior Angles in Geometric Figures

1. What term describes a pair of angles formed by the intersection of two straight lines that share a common vertex but do not share any common sides?

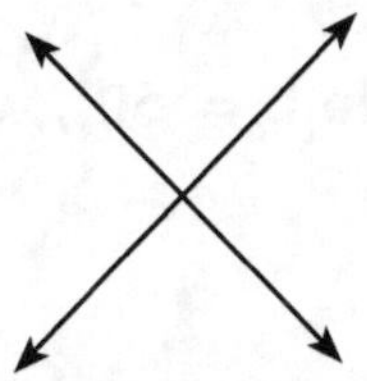

Ⓐ Supplementary Angles
Ⓑ Complementary Angles
Ⓒ Horizontal Angles
Ⓓ Vertical Angles

2. If a triangle has two angles with measures that add up to 100 degrees, what must the measure of the third angle be?

Ⓐ 180 degrees
Ⓑ 100 degrees
Ⓒ 80 degrees
Ⓓ 45 degrees

3.

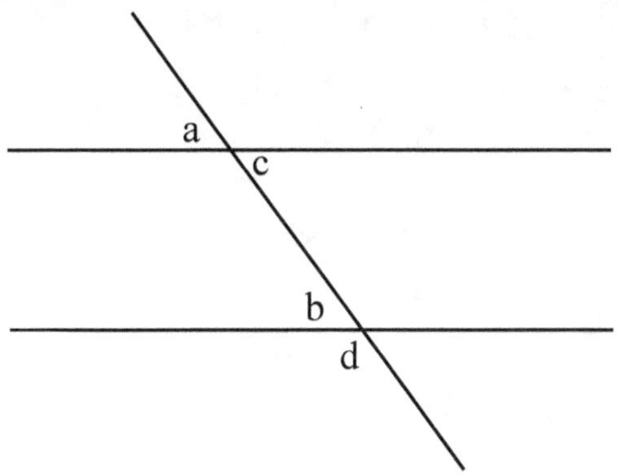

The figure shows two parallel lines intersected by a third line. If a = 55°, what is the value of b?

Ⓐ 35°
Ⓑ 45°
Ⓒ 55°
Ⓓ 125°

4.

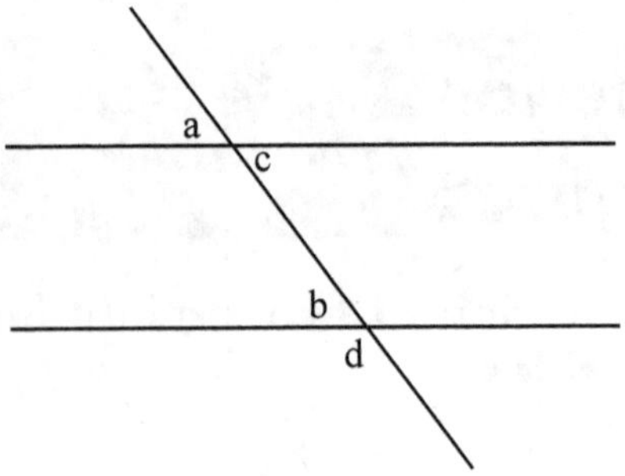

The figure shows two parallel lines intersected by a transversal. If angle b = 60°, what is the measure of angle c?

Ⓐ 30°
Ⓑ 60°
Ⓒ 90°
Ⓓ 120°

5.

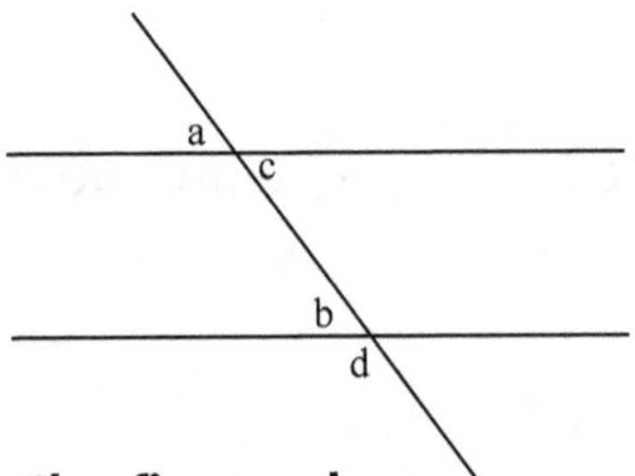

The figure shows two parallel lines intersected by a transversal. If angle d = 130°, what is the measure of angle a?

Ⓐ 30°
Ⓑ 40°
Ⓒ 50°
Ⓓ 130°

6.

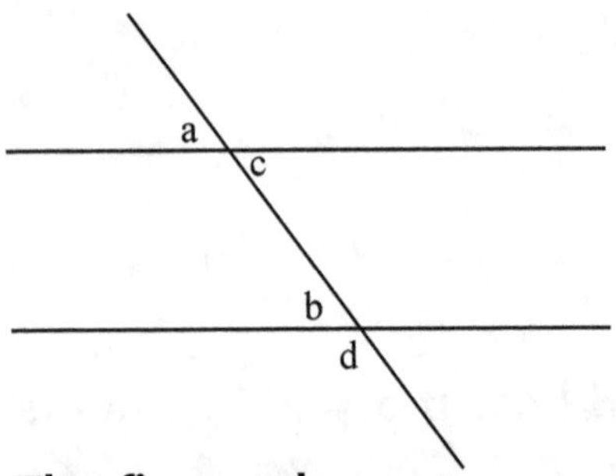

The figure shows two parallel lines intersected by a third line. Which of the following angles are equal in measure?

Ⓐ a and b only
Ⓑ a and c only
Ⓒ b and c only
Ⓓ a, b, and c

7. Two angles in a triangle measure 65° each. What is the measure of the third angle in the triangle?

Ⓐ 25°
Ⓑ 50°
Ⓒ 65°
Ⓓ 130°

8. 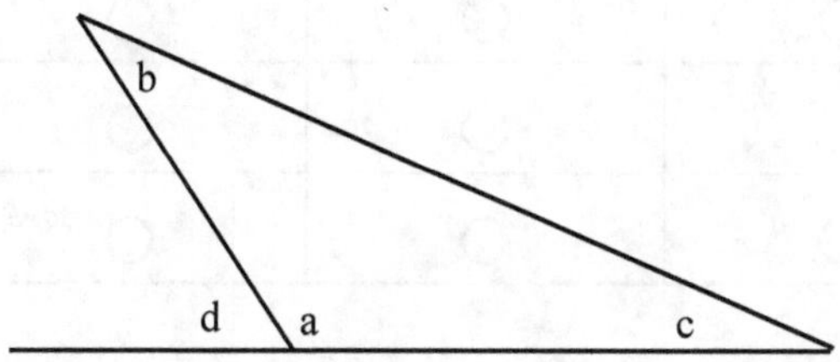

If b = 40° and c = 30°, what is the measure of d?

Ⓐ 35°
Ⓑ 70°
Ⓒ 110°
Ⓓ 145°

9. In right triangle ABC, Angle C is the right angle. Angle A measures 70°. Find the measure of the exterior angle at angle C.

Ⓐ 180°
Ⓑ 90°
Ⓒ 110°
Ⓓ 160°

10. If two parallel lines are cut by a transversal, the alternate interior angles are _____.

Ⓐ supplementary
Ⓑ complementary
Ⓒ equal in measure
Ⓓ none of the above

11. Match each polygon with the correct sum of its interior angles.

	2520	1080	1440	4140	540
Decagon	○	○	○	○	○
16-gon	○	○	○	○	○
Pentagon	○	○	○	○	○
25-gon	○	○	○	○	○
Octagon	○	○	○	○	○

12. Observe the figure given below. ∠3 and ∠7 are what type of angles?

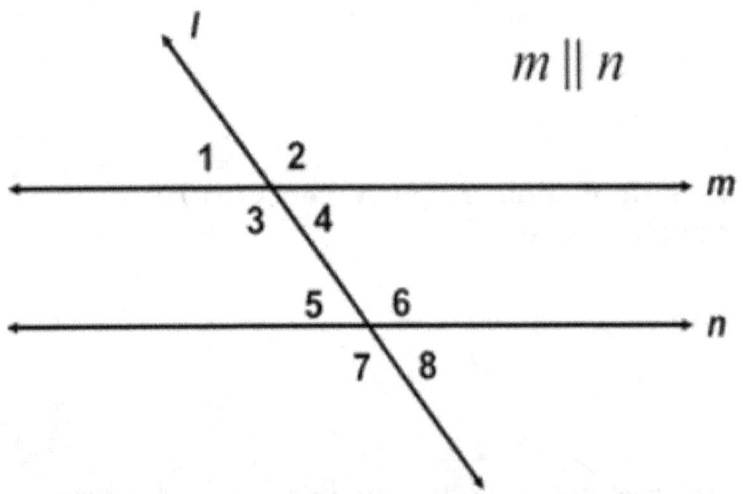

∠3 and ∠7 are what type of angles? Write your answer in the box given below.

CHAPTER 4 → Lesson 8: Verifying the Pythagorean Theorem

1. Which of the following could be the lengths of the sides of a right triangle?

Ⓐ 1, 2, 3
Ⓑ 2, 3, 4
Ⓒ 3, 4, 5
Ⓓ 4, 5, 6

2. A triangle has sides 8 cm long and 15 cm long, with a 90° angle between them. What is the length of the third side?

Ⓐ 7 cm
Ⓑ 17 cm
Ⓒ 23 cm
Ⓓ 289 cm

3. Find the length of side c, rounded to the nearest tenth.

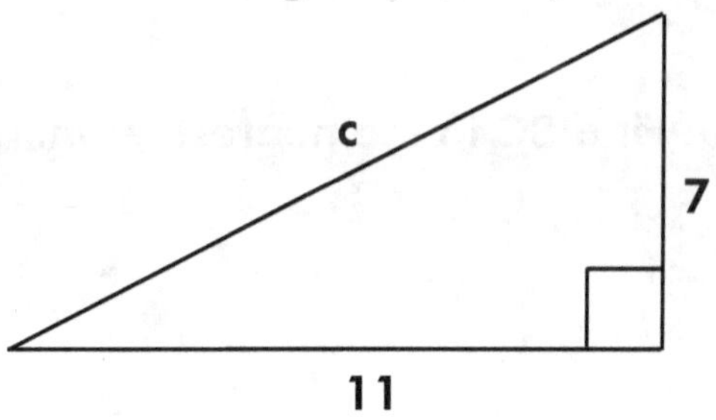

Ⓐ 8.5
Ⓑ 8.8
Ⓒ 13.0
Ⓓ 19.3

4. A square has sides 5 inches long. What is the approximate length of a diagonal of the square?

Ⓐ 5 inches
Ⓑ 6 inches
Ⓒ 7 inches
Ⓓ 8 inches

5. **Which of the following INCORRECTLY completes this statement of the Pythagorean theorem?**
In a right triangle with legs of lengths a and b and hypotenuse of length c, ...

Ⓐ $a^2 + b^2 = c^2$
Ⓑ $c^2 - a^2 = b^2$
Ⓒ $c^2 - b^2 = a^2$
Ⓓ $a^2 + c^2 = b^2$

6. **A Pythagorean triplet is a set of three positive integers a, b, and c that satisfy the equation $a^2 + b^2 = c^2$. Which of the following is a Pythagorean triple?**

Ⓐ a = 3, b = 6, c = 9
Ⓑ a = 6, b = 9, c = 12
Ⓒ a = 9, b = 12, c = 15
Ⓓ a = 12, b = 15, c = 18

7. **If an isosceles right triangle has legs of 4 inches each, find the length of the hypotenuse.**

Ⓐ Approximately 6 in.
Ⓑ Approximately 5 in.
Ⓒ Approximately 4 in.
Ⓓ Approximately 3 in.

8. **In triangle ABC, angle C = 90°, AC = 4 and AB = 10. Find BC to the nearest tenth.**

Ⓐ 9.5
Ⓑ 9.2
Ⓒ 8.9
Ⓓ 8.5

9. **In triangle ABC, angle C = 90°, AB = 35, and BC = 28. Find AC.**

Ⓐ 23
Ⓑ 22
Ⓒ 21
Ⓓ 20

10. **The diagonal of a square is 25. Find the approximate side lengths.**

Ⓐ 15
Ⓑ 16
Ⓒ 17
Ⓓ 18

11. You have a right triangle with leg lengths of 6 and 8. What is the length of the hypotenuse?

$6^2 + 8^2 = C^2$
$100 = C^2$
$\sqrt{100} = C$
C=?

Ⓐ 4
Ⓑ 10
Ⓒ 36
Ⓓ 64

12. Which equation would you use to solve for the missing side of the triangle pictured below?

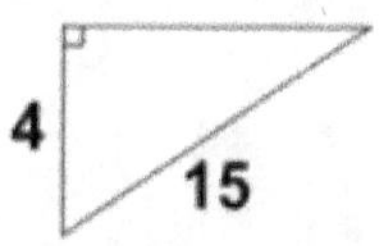

Ⓐ $4^2+15^2=x^2$
Ⓑ $4^2+x^2=15^2$
Ⓒ $x^2+15^2=4^2$

CHAPTER 4 → Lesson 9: Pythagorean Theorem in Real-World Problems

1. The bottom of a 17-foot ladder is placed on level ground 8 feet from the side of a house as shown in the figure below. Find the vertical height at which the top of the ladder touches the side of the house.

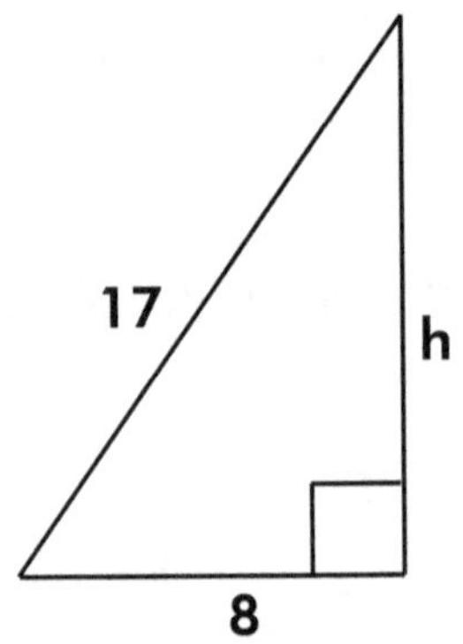

Ⓐ h = 9 feet
Ⓑ h = 12 feet
Ⓒ h = 15 feet
Ⓓ h = 18 feet

2. Which of the following equations could be used to find the value of w?

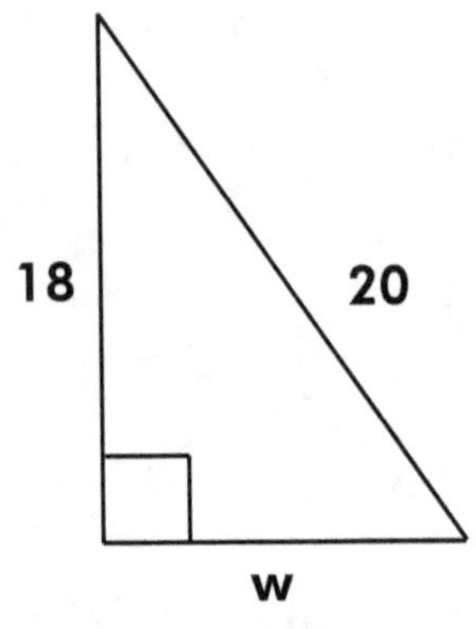

Ⓐ $w^2 + 18^2 = 20^2$
Ⓑ $18^2 - w^2 = 20^2$
Ⓒ $20^2 + 18^2 = w^2$
Ⓓ $w + 18 = 20$

3. **John has a chest where he keeps his antiques. What is the measure of the diagonal (d) of John's chest with the height (c) = 3ft, width (b) = 3ft, and length (a) = 5ft.?**

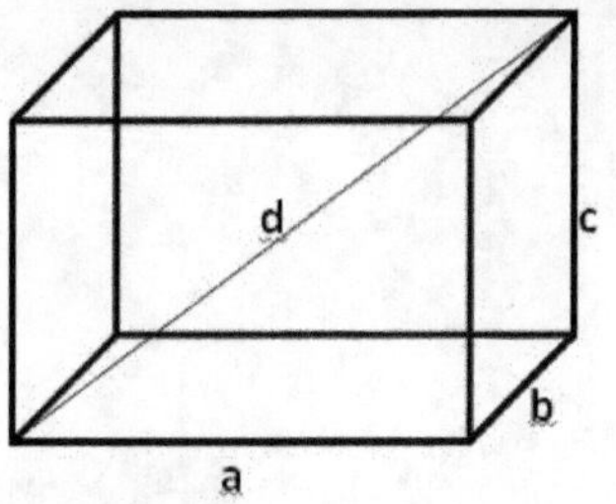

Ⓐ $\sqrt{42}$ ft²
Ⓑ $\sqrt{44}$ ft²
Ⓒ $\sqrt{34}$ ft
Ⓓ $\sqrt{43}$ ft

4. **Mary has a lawn that has a width (a) of 30 yards, and a length (b) of 40 yards. What is the measurement of the diagonal (c) of the lawn?**

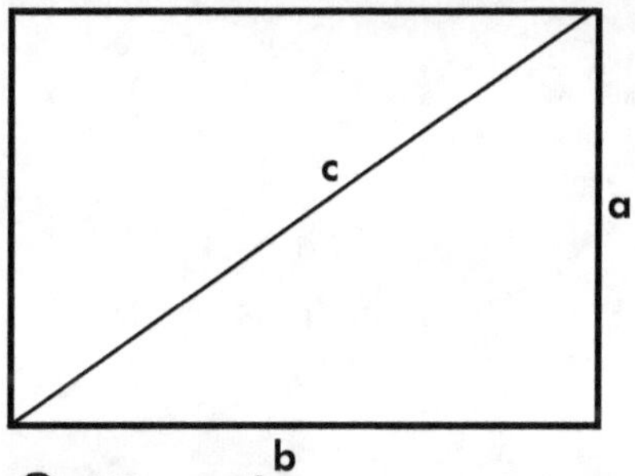

Ⓐ 50 yards
Ⓑ 49 yards
Ⓒ 51 yards
Ⓓ 50 yards^2

5. **A construction company needed to build a sign with the width (a) of 9 ft, and a length (b) of 20 ft. What will be the approximate measurement of the diagonal (c) of the sign?**

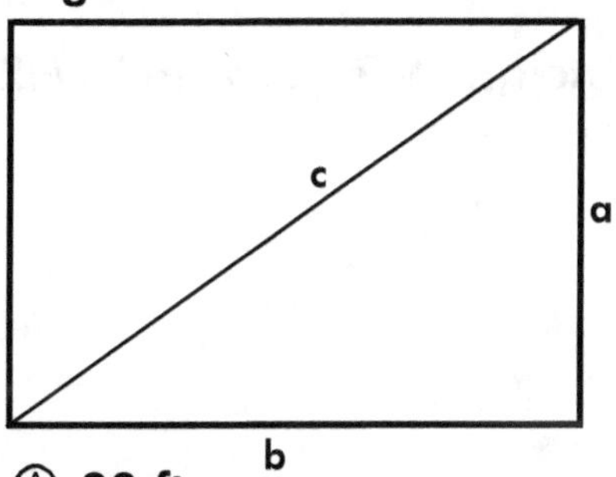

Ⓐ 23 ft
Ⓑ 22 ft
Ⓒ 20 ft
Ⓓ 19 ft

6. An unofficial baseball diamond is measured to be 50 yards wide. What is the approximate measurement of one side (a) of the diamond?

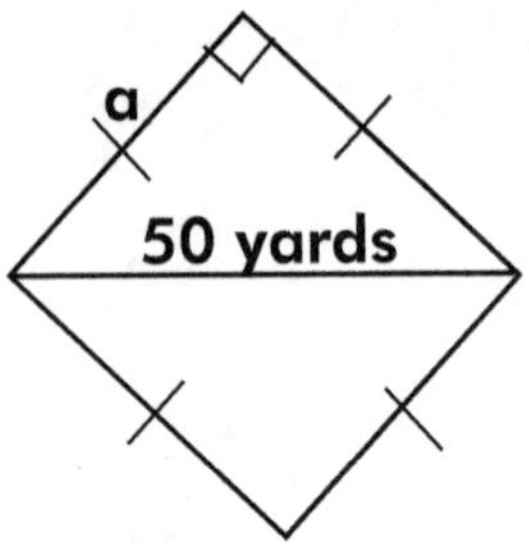

Ⓐ 34 ft
Ⓑ 35 yards
Ⓒ 35 ft
Ⓓ 36 yards

7. The neighborhood swimming pool is 20 ft wide and 30 ft long. What is the approximate measurement of the diagonal (d) of the base of the pool?

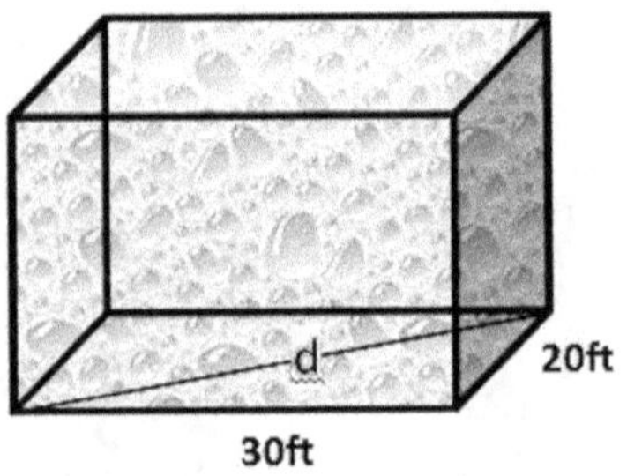

Ⓐ 36 ft
Ⓑ 35 ft
Ⓒ 34 ft
Ⓓ 34 yards

8. To get to her friend's house, a student must walk 20 feet to the corner of their streets, turn left and walk 15 feet to her friend's house.
How much shorter would it be if she could cut across a neighbor's yard and walk a straight line from her house to her friend's house?

Ⓐ 5 feet shorter
Ⓑ 10 feet shorter
Ⓒ 25 feet shorter
Ⓓ 35 feet shorter

9. Your school principal wants the custodian to put a new flag up on the flagpole. If the flagpole is 40 feet tall and they have a 50 foot ladder, approximately how far from the base of the pole can he place the base of his ladder in order to accomplish the task?

Ⓐ Up to 10 ft away
Ⓑ Up to 20 ft away
Ⓒ Up to 30 ft away
Ⓓ Up to 40 ft away

10. Your kite is at the end of a 50 ft string. You are 25 ft from the outside wall of a building that you know makes a right angle with the ground.
How high is the kite approximately?

Ⓐ Approximately 39 ft high
Ⓑ Approximately 40 ft high
Ⓒ Approximately 42 ft high
Ⓓ Approximately 43 ft high

11. Match each word problem with the correct equation you would use to solve it.

	$9^2+x^2=15^2$	$40^2+38^2=x^2$	$9^2+15^2=x^2$	$x^2+38^2=40^2$
One house is 15 miles due north of the park. Another house is 9 miles due east of the park. How far apart are the houses from each other?	○	○	○	○
The foot of a ladder is put 9 feet from the wall. If the ladder is 15 feet long how high up the building will the ladder reach?	○	○	○	○
If you drive your car 40 miles south and then 38 miles east, how far would the shortest route be from your starting point?	○	○	○	○
The diagonal of a TV is 40 inches. The TV is 38 inches long. How tall is the TV?	○	○	○	○

12. Fill in the missing side lengths needed to solve each word problem. Round your answers to the nearest tenth if necessary.

WORD PROBLEM	LEG(a)	LEG(b)	HYPOTENUSE(c)
Find the height of a pyramid whose slant height is 26 cm and base length is 48 cm	24		
Find the base length of a pyramid whose height is 8 in and slant height 17 in.		8	17
The foot of a ladder is put 5 feet from the wall. If the top of the ladder is 10 feet from the ground, how long is the ladder?		10	

13. On a bike ride you start at your house and ride your bike 3.5 miles north and then 1.25 miles east. How far are you directly from your house? Select the equation and solution that correctly solve the problem.

Ⓐ $\sqrt{3.5^2+1.25^2}$ = 22.56 mi
Ⓑ $\sqrt{3.5^2-1.25^2}$ = 3.34 mi
Ⓒ $\sqrt{3.5^2+1.25^2}$ = 3.72 mi

Date of Completion:____________ Score:____________

CHAPTER 4 → Lesson 10: Pythagorean Theorem & Coordinate System

1. A robot begins at point A, travels 4 meters west, then turns and travels 7 meters south, reaching point B. What is the approximate straight-line distance between points A and B?

   Ⓐ 8 meters
   Ⓑ 9 meters
   Ⓒ 10 meters
   Ⓓ 11 meters

2. What is the distance between the points (1, 3) and (9, 9)?

   Ⓐ 6 units
   Ⓑ 8 units
   Ⓒ 10 units
   Ⓓ 12 units

3. Find the distance (approximately) between Pt A (2, 7) and Pt B (-2, -7).

   Ⓐ 14.0
   Ⓑ 14.6
   Ⓒ 18.0
   Ⓓ 13.4

4. Find the distance (approximately) between Pt P (5, 3) and the origin (0, 0).

   Ⓐ 4.0
   Ⓑ 5.1
   Ⓒ 5.8
   Ⓓ 8.0

5. Find the distance (approximately) between the points A (11, 12) and B (7, 8).

   Ⓐ 4.0
   Ⓑ 5.3
   Ⓒ 5.7
   Ⓓ 8.0

6. Is it closer to go from Pt A (4, 6) to Pt B (2, -4) or Pt A to Pt C (-5, 2)?

   Ⓐ A to B
   Ⓑ A to C
   Ⓒ Neither, they are both the same distance.
   Ⓓ Not enough information.

7. Paul lives 50 yards east and 40 yards south of his friend, Larry. If he wants to shorten his walk by walking in a straight line from his home to Larry's, how far (approximately) will he walk?

Ⓐ 64 yards
Ⓑ 62 yards
Ⓒ 60 yards
Ⓓ 58 yards

8. In a coordinate plane, a point P (7,8) is rotated 90° clockwise around the origin and then reflected across the vertical axis. Find the distance (approximately) between the original point and the final point.

Ⓐ 14 units
Ⓑ 21 units
Ⓒ 24 units
Ⓓ None of the above

9. You are using a coordinate plane to sketch out a plan for your vegetable garden. Your garden will be a rectangle 15 ft wide and 20 ft long. You want a square in the center with 3 ft sides to be reserved for flowers. If the garden is plotted on the coordinate plane with the southwest corner at the origin, what are the coordinates of the center of the flower garden? (Assume that the longer side is placed on the y-axis)

Ⓐ (15, 10)
Ⓑ (7.5, 10)
Ⓒ (-15, 10)
Ⓓ (-7.5, 10)

10. You are using a coordinate plane to sketch out a plan for a vegetable garden. The garden will be a rectangle 15 feet wide and 20 feet long. You want a square in the center with 3 feet sides to be reserved for flowers. The garden is plotted in the coordinate grid so that the southwest corner is placed at the origin. Find the length of the diagonal (approximately) of the flower bed.

Ⓐ 4.0 feet
Ⓑ 4.2 feet
Ⓒ 5.0 feet
Ⓓ 5.2 feet

11. Match the ordered pairs with the approximate distance between them.

	10.8	12.2	13	14.8
(6, 5) and (-4, 9)	○	○	○	○
(-8, 0) and (5, -7)	○	○	○	○
(-4, -9) and (6, -2)	○	○	○	○
(5, 4) and (12, 15)	○	○	○	○

12. Given below are the coordinates of the hypotenuse of right-angled triangles. Fill in the missing values with the lengths of the sides. Round to the nearest hundredth if necessary.

ORDERED PAIRS	LEG LENGTH	LEG LENGTH	HYPOTENUSE
(6,2),(0,-6)	6		
(-3,-1),(-4,0)	1	1	
(-2,3),(-1,7)			4.12

CHAPTER 4 → Lesson 11: Finding Volume: Cone, Cylinder, & Sphere

1. What is the volume of a sphere with a radius of 6?

Ⓐ 72π
Ⓑ 144π
Ⓒ 216π
Ⓓ 288π

2. A cone has a height of 9 and a base whose radius is 4. Find the volume of the cone.

Ⓐ 18π
Ⓑ 36π
Ⓒ 48π
Ⓓ 72π

3. What is the volume of a cylinder with a radius of 5 and a height of 3?

Ⓐ 30π
Ⓑ 45π
Ⓒ 75π
Ⓓ 120π

4. A round ball has a diameter of 10 inches. What is the approximate volume of the ball? Use $\pi = 3.14$

Ⓐ 130 cubic inches
Ⓑ 260 cubic inches
Ⓒ 390 cubic inches
Ⓓ 520 cubic inches

5. A cylindrical can has a height of 5 inches and a diameter of 4 inches. What is the approximate volume of the can? Use $\pi = 3.14$

Ⓐ 20 cubic inches
Ⓑ 60 cubic inches
Ⓒ 100 cubic inches
Ⓓ 200 cubic inches

6. A cylinder and a cone have the same radius and the same volume. How do the heights compare?

Ⓐ The height of the cylinder is 3 times the height of the cone.
Ⓑ The height of the cylinder is 2 times the height of the cone.
Ⓒ The height of the cone is 2 times the height of the cylinder.
Ⓓ The height of the cone is 3 times the height of the cylinder.

7. Which of the following has the greatest volume?

Ⓐ A sphere with a radius of 2 cm
Ⓑ A cylinder with a height of 2 cm and a radius of 2 cm
Ⓒ A cone with a height of 4 cm and a radius of 3 cm
Ⓓ All three volumes are equal

8. Which of the following has the greatest volume?

Ⓐ A sphere with a radius of 3 cm
Ⓑ A cylinder with a height of 4 cm and a radius of 3 cm
Ⓒ A cone with a height of 3 cm and a radius of 6 cm
Ⓓ All three volumes are equal

9. If a sphere and a cone have the same radius r, and the cone has a height of 4, find the ratio of the volume of the sphere to the volume of the cone.

Ⓐ r : 1
Ⓑ 1 : r
Ⓒ r : 4
Ⓓ 4 : r

10. Which of the following describes the relationship between the volumes of a cone and cylinder with the same radii and the same height.

Ⓐ The volume of the cylinder is three times that of the cone.
Ⓑ The volume of the cylinder is 1/3 that of the cone.
Ⓒ The volume of the cylinder is 4/3 that of the cone.
Ⓓ None of the above.

11. **Find the volume of the figure below. Use pi = 3.14. Write your answer in the box below.**

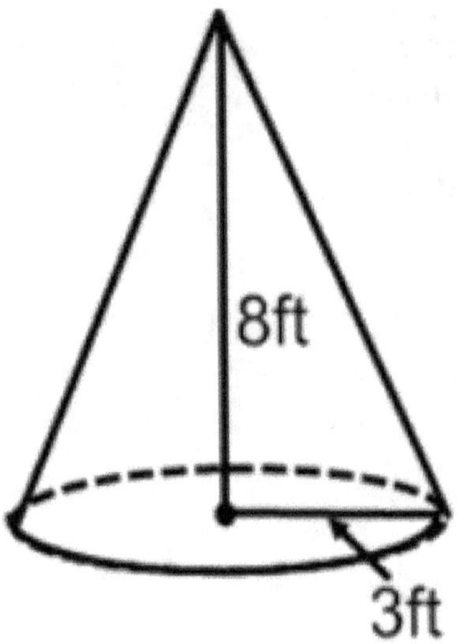

12. **Select the equation you would use to find the volume of the sphere pictured.**

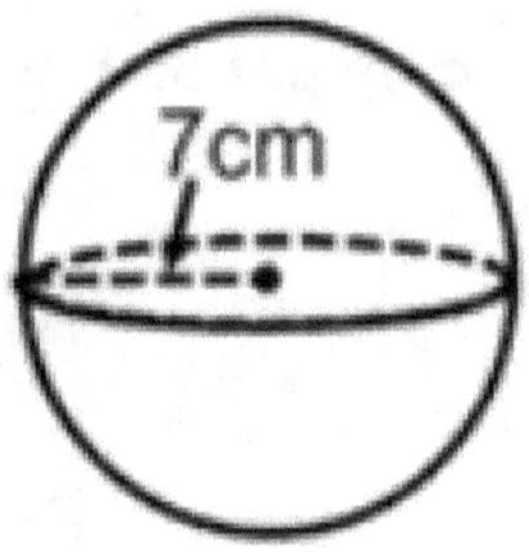

Ⓐ $4 \times \pi \times 7^2$

Ⓑ $\frac{4}{3} \times \pi \times 7^3$

Ⓒ $\frac{4}{3} \times \pi \times 7^2$

End of Geometry

ANSWER KEY AND DETAILED EXPLANATION

Chapter 4: Geometry

Lesson 1: Transformations of Points and Lines

Question No.	Answer	Detailed Explanation
1	D	If you rotate (4,3) 90° clockwise, it will move from quadrant I to quadrant IV. x becomes 3 and y becomes -4.
2	D	There is not enough information provided. All three of the transformations listed would keep the length of the segment the same.
3	A	x becomes -x and y becomes -y as the point rotates from quadrant I to quadrant III.
4	B	This is a reflection across the y-axis.
5	C	A simple slide to the left within the coordinate plane is an example of a translation.
6	B	This is a rotation, since the segment is still the same length, it is just oriented differently.
7	C	This is a reflection across the x-axis, since the x-coordinates are remaining unchanged and the y-coordinates are switching sign.
8	A	This would be a rotation through 90° clockwise.
9	D	This is either a reflection in the x-axis, a translation downward, or a 180° rotation about the origin.
10	D	Reflection across the y-axis will change the sign of x but keep the same y value.

Question No.	Answer	Detailed Explanation
11		If a rotation, translation, or reflection takes place, the shape will still be the same size. Therefore, third statement is true. Also, if reflecting over the y-axis, you take the opposite of your x-coordinate and keep your y-coordinate the same. Therefore, second statement is true. During translation, you will either add or subtract to the x-coordinate or y-coordinate depending on which direction you are moving. If you are moving to the left, you will subtract from the x-coordinate. If you move to the right, you will add to the x-coordinate. If you move up, you will add to the y-coordinate and if you move down, you will subtract from the y-coordinate. Applying these rules of translation, we see that fourth statement is true and first statement is false. In the first case, the coordinates of the point A (3, 2), after translation by 10 units down will be A' (3, -8).
12	x-3; y+4	In this case since you are translating left, you will need to put a subtraction sign in the first box. In the second box, since you are moving up, you will need to put an addition sign in there.
13	C	The third picture represents a rotation because you can tell that point A has rotated and the figure has changed its orientation.

Lesson 2: Transformations of Angles

Question No.	Answer	Detailed Explanation
1	D	Corresponding angles are congruent under rigid transformations. Since triangle BDE is a reflection image of triangle ABC, and angle C corresponds to angle E, the two angles are equal in measure.
2	D	Corresponding angles are congruent under rigid transformations. Since triangle DEF is a rotated image of triangle ABC, and angle C corresponds to angle F, the two angles are equal in measure.
3	B	Since figure ZYXWV is a rotated image of figure ABCDE, this transformation will prove angle ABC is congruent to angle XYZ.
4	A	Since figure XYZVW is a reflected image of figure ABCDE, this transformation will prove angle ABC is congruent to angle XYZ.
5	B	Correct Answer: Option (B) Corresponding angles are congruent under rigid transformations. Since triangle DEF is a rotated image of triangle ABC, and angle B in ABC corresponds to angle D in DEF, the two angles are equal in measure.
6	D	All figures that undergo rigid transformations maintain their congruency, which includes side measures, angles, and areas. Thus, D is the correct answer.
7	D	Translation, rotation and reflection do not alter the size of a shape so the two triangles can not be congruent using these transformations.
8	D	Correct Answer: Option (D) All of the logos shown are results of only reflection, rotation, or translation. Therefore, all these logos would be acceptable.
9	B	All figures that undergo rigid transformations have corresponding angles that maintain congruency. Since angle AOB corresponds to angle XYZ, the two are congruent and therefore both 45 degrees.
10	A	Correct Answer: Option (A) Corresponding angles are congruent under rigid transformations. Since figure ABCDEF is a reflected image of figure MNLKJI, and angle ABC corresponds to angle MNL, the two angles are congruent and therefore both measure 110 degrees.

<table>
<tr><th>Question No.</th><th>Answer</th><th>Detailed Explanation</th></tr>
<tr><td>11</td><td></td><td>
<table>
<tr><th></th><th>30°</th><th>60°</th><th>90°</th></tr>
<tr><td>Translation</td><td>●</td><td>○</td><td>○</td></tr>
<tr><td>Reflection</td><td>●</td><td>○</td><td>○</td></tr>
<tr><td>Rotation</td><td>●</td><td>○</td><td>○</td></tr>
<tr><td>Dilation</td><td>●</td><td>○</td><td>○</td></tr>
</table>
Angle measures will stay the same no matter what transformation they undergo.</td></tr>
<tr><td>12</td><td>E</td><td>In this figure, you can tell that the pre-image is triangle ABC. When it was rotated, you can see that the angle measures stayed the same. Thus, you are looking for the obtuse angle. In this case, angle E corresponds to angle B in the pre-image.</td></tr>
<tr><td>13</td><td>D</td><td>If you rotate the figure 90 degrees counterclockwise, you will see that angle B lines up with angle D, matching angle A and angle E, leaving the correct answer as angle C and angle F.</td></tr>
</table>

Lesson 3: Transformations of Parallel Lines

Question No.	Answer	Detailed Explanation
1	B	This is a translation, since the two segments are still oriented in the same way. They were just shifted downward in the coordinate grid.
2	D	This transformation is not a translation, since the line segments are now directed differently. It could be a reflection or a rotation, since both would change the way the segments are facing.
3	D	This transformation is not a translation, since the line segments are now directed differently. It could be a reflection or a rotation, since both would change the way the segments are facing.
4	D	When a line is translated along a vector, the resulting line is parallel.
5	D	When a line is translated along a vector, the resulting line is parallel.
6	C	When a line is translated along a vector, the resulting line is parallel.
7	B	Given a line and a point, there is one and only one line that can be drawn through the point that is parallel to the original line.
8	B	Since ABCD is a rectangle, its opposite sides are parallel by definition. Translations always map parallel lines to parallel lines, therefore $\overline{XY} \parallel \overline{WZ}$
9	D	Vertical parallel lines rotated 180 degrees will result in another set of parallel vertical lines. Option A is not the answer because it is the result of a translation, not a rotation.
10	A	A 180-degree rotation will always map a segment to a parallel segment.

<table>
<tr><th>Question No.</th><th>Answer</th><th>Detailed Explanation</th></tr>
<tr><td>11</td><td></td><td>
<table>
<tr><th></th><th>Lengths of sides</th><th>Angle Measures</th><th>Parallel Sides on Figure</th></tr>
<tr><td>Translation</td><td>✓</td><td>✓</td><td>✓</td></tr>
<tr><td>Reflection</td><td>✓</td><td>✓</td><td>✓</td></tr>
<tr><td>Rotation</td><td>✓</td><td>✓</td><td>✓</td></tr>
<tr><td>Dilation</td><td></td><td>✓</td><td>✓</td></tr>
</table>
All of the transformations will preserve the angle measures. Parallel sides of the figure is' to 'Parallel sides of the figure are' to agree in number. Lengths of the sides will change when a figure undergoes a dilation. In other three transformations, lengths of sides is also preserved.</td></tr>
<tr><td>12</td><td>-1/3</td><td>Figure ABCD was translated. If a figure is translated it keeps parallel sides parallel. So in this case AC has the slope of -1/3. Thus meaning that A'C' will also have a slope of -1/3.</td></tr>
<tr><td>13</td><td>B</td><td>In this case the middle answer will correctly show a reflection over the y-axis. It is also demonstrating that since the lines were parallel to start, they will still be parallel after the reflection.</td></tr>
</table>

Lesson 4: Transformations of Congruency

Question No.	Answer	Detailed Explanation
1	B	Of the choices given, the wings of a butterfly are nearly always congruent; i.e. the same shape and the same size.
2	B	Reflection preserves congruence.
3	D	We can't tell if it was a reflection across the y-axis or a translation.
4	C	It shows a translation.
5	A	The object in quadrant III could be a reflection of the object in quadrant II. Therefore, option (A) is the correct answer.
6	A	Two congruent objects are identical and consequently have the same dimensions.
7	D	Only dilation does not preserve congruency.
8	B	A figure formed by rotation, reflection or translation or any combination of these three transformations will be congruent to the original.
9	D	If two triangles are congruent, then they have congruent sides, congruent angles, and equal areas. And triangles have three sides. Therefore option (D) is the correct answer.
10	B	Because the letter O doesn't have a right and left that are different and a top and bottom that are different, it looks the same after this sequence of transformations.
11		This is showing two figures that have been obtained from sliding one onto the other. If you can get from the pre-image to the image through rigid transformations, then the two figures are congruent. If they are congruent, they are also similar.
12	IJKL	If you apply the transformation to the original figure you will see that quadrilater ABCD will land directly on top of quadrilateral IJKL. This will also show that the quadrilaterals will be congruent.

Lesson 5: Analyzing Transformations

<table>
<tr><th>Question No.</th><th>Answer</th><th>Detailed Explanation</th></tr>
<tr><td>1</td><td>C</td><td>A translation slides a figure or point without any alterations in size or shape.</td></tr>
<tr><td>2</td><td>A</td><td>Triangle A has been rotated to produce triangle B.</td></tr>
<tr><td>3</td><td>B</td><td>Triangle A has been reflected to produce triangle B.</td></tr>
<tr><td>4</td><td>D</td><td>The correct answer choice is D. Triangle A cannot be transformed into Triangle B through rotation, reflection, or translation in a single step.</td></tr>
<tr><td>5</td><td>C</td><td>Triangle A has been translated to produce triangle B.</td></tr>
<tr><td>6</td><td>B</td><td>Triangle A has been reflected to produce triangle B.</td></tr>
<tr><td>7</td><td>A</td><td>Triangle A has been reflected and translated to produce Triangle B.</td></tr>
<tr><td>8</td><td>D</td><td>Dilation does not preserve congruency.</td></tr>
<tr><td>9</td><td>B</td><td>A point (x, y) reflected in the vertical axis becomes (-x, y).
A point (- x, y) reflected in the horizontal axis would become (-x, -y).</td></tr>
<tr><td>10</td><td>A</td><td>When translating a point (x, y) 3 units to the left, it will be located at (x-3, y).</td></tr>
<tr><td>11</td><td></td><td>
<table>
<tr><th></th><th>A(4, -2)</th><th>A(-2, 1)</th><th>A(-4, -2)</th></tr>
<tr><td>Translation (x+2, y-1)</td><td></td><td>●</td><td></td></tr>
<tr><td>Rotation 180°</td><td>●</td><td></td><td></td></tr>
<tr><td>Reflection over x-axis</td><td></td><td></td><td>●</td></tr>
</table>
By applying each of the transformations to the pre-image, you can determine the ordered pair for point A. If you translate the original point (-4,2) 2 to the right and one down you end up with (-2,1). If it is a 180 degree rotation you take the opposite of your x-coordinate and the opposite of the y-coordinate. Thus ending up with (4,-2). Finally, if it is a reflection over the x-axis your x-coordinate stays the same and you take the opposite of your y-coordinate. Thus ending up with (-4,-2).</td></tr>
</table>

Question No.	Answer	Detailed Explanation
12	2	Since the triangle became larger from the original to the image, you know the scale factor must be greater than one. To find the scale factor you take one of the side lengths from the image and put it over the corresponding side length from the pre-image (or calculate the lengths using distance formula). In this case I used Side BC and EF. Length of EF is 12 and the length of BC is 6. length of EF / length of BC = 12 / 6 = 2. Therefore, Scale factor is 2.

Lesson 6: Transformations & Similarity

Question No.	Answer	Detailed Explanation
1	D	A dilation could transform Triangle A to Triangle B because it shrinks or enlarges a figure.
2	C	The large object has been translated and dilated to produce the smaller object.
3	A	Since the shape of the object has changed, it was not produced by any of the four transformations.
4	B	In the second choice, the larger object was reflected across the horizontal axis and dilated.
5	C	In order to have two similar objects, we must have a dilation.
6	A	One condition of similarity is that all dimensions are proportional.
7	C	The first transformation rotated the point 90° clockwise and the second translated it 3 units to the left.
8	A	AB : WX = 5 : 10 = 1 : 2 BC : XY = 3 : 6 = 1 : 2 AC : WY = 6 : 12 = 1 : 2
9	B	Sides of rectangles A, C and D are proportional. If we take rectangle A and C, ratio of their lengths = 1 : 2, ratio of their widths = 2 : 4 = 1 : 2 = ratio of their lengths. Therefore rectangles A and C are similar. If we take rectangle A and D, ratio of their lengths = 1 : 3, ratio of their widths = 2 : 6 = 1 : 3 = ratio of their lengths. Therefore rectangles A and D are similar. Since rectangles C and D are both similar to rectangle A, they are similar to each other. If we take rectangle A and B, ratio of their lengths = 1 : 2, ratio of their widths = 2 : 3. Ratio of lengths is not the same as ratio of widths. Therefore A and B are not similar.

<table>
<tr><th>Question No.</th><th>Answer</th><th>Detailed Explanation</th></tr>
<tr><td>10</td><td>C</td><td>AB : WX = AC : WY
9 : 27 = 14 : WY
3 : 9 = 14 : WY
3(WY) = 9(14)
WY = 42</td></tr>
<tr><td>11</td><td></td><td><table>
<tr><th></th><th>Used</th><th>Similar only</th><th>Congruent</th></tr>
<tr><td>Translation</td><td>✓</td><td></td><td>✓</td></tr>
<tr><td>Rotation</td><td></td><td></td><td>✓</td></tr>
<tr><td>Reflection</td><td></td><td></td><td>✓</td></tr>
<tr><td>Dilation</td><td>✓</td><td>✓</td><td></td></tr>
</table>
You can translate to line them up. Whenever you translate you keep the same shape and size, so the figures are congruent. Next, you know you have to dilate because the sizes are different. Whenever you dilate it keeps the same shape, but changes the size. Thus keeping the figures similar.</td></tr>
<tr><td>12</td><td>8</td><td>Since the two figures are similar, you can either figure out the scale factor between them or set up a proportion. If you set up a proportion it would look like 3/6 = 4/h. Then you cross multiply and solve. Otherwise, if you look at similar parts, you can figure out how you get from one to the other. In this case you will multiply the left figure by 2 to get the right figure. So in this case h=8.</td></tr>
</table>

Lesson 7: Interior & Exterior Angles in Geometric Figures

Question No.	Answer	Detailed Explanation
1	D	The statement in the problem is the definition of vertical angles.
2	C	$a + b + c = 180$ $100 + c = 180$ $c = 80°$
3	C	If two parallel lines are cut by a transversal, the corresponding angles are congruent.
4	B	If two parallel lines are cut by a transversal, the alternate interior angles are congruent.
5	C	If two parallel lines are cut by a transversal, the exterior angles on the same side of the transversal are supplementary.
6	D	If two parallel lines are cut by a transversal the alternate interior angles are congruent; so the measures of angle b and angle c are equal. If two straight lines intersect, the vertical angles are congruent; so the measures of angle a and angle c are equal. That makes the measures of angles a, b, and c all equal.
7	B	$a + b + c = 180°$ $65° + 65° + c = 180°$ $130° + c = 180°$ $c = 50°$
8	B	$a + b + c = 180°$ $a + 40° + 30° = 180°$ $a = 110°$ $a + d = 180°$ $110° + d = 180°$ $d = 70°$ Also, the measure of an exterior angle of a triangle is equal to the sum of the measures of the two non-adjacent interior angles.
9	B	An interior angle of a triangle and its exterior angle are supplementary.
10	C	If two parallel lines are cut by a transversal, the measures of the alternate interior angles are equal.

<table>
<tr><th>Question No.</th><th>Answer</th><th>Detailed Explanation</th></tr>
<tr><td>11</td><td></td><td>

	2520	1080	1440	4140	540
Decagon			●		
16-gon	●				
Pentagon					●
25-gon				●	
Octagon		●			

To find the sum of the interior angles of a polygon, you use the formula the number of sides minus 2 times 180. If n is the number of sides (or interior angles), then the sum of the interior angles of the polygon is (n - 2) x 180°.

(1) For Decagon, n = 10, sum of the interior angles = (10 - 2) x 180° = 1440°

(2) For 16-gon, n = 16, sum of the interior angles = (16 - 2) x 180° = 2520°

(3) For Pentagon, n = 5. sum of the interior angles = (5 - 2) x 180° = 540°

(4) For 25-gon, n = 25, sum of the interior angles = (25 - 2) x 180° = 4140°

(5) For Octagon, n = 8, sum of the interior angles = (8 - 2) x 180° = 1080°
</td></tr>
<tr><td>12</td><td>corresponding</td><td>These angles are on the same side of their parallel lines and the same side of the transversal. This means that these angles are Corresponding angles. Another way of identifying corresponding angles is : they are in the same location at each point of intersection of transversal and parallel lines.</td></tr>
</table>

Lesson 8: Verifying the Pythagorean Theorem

Question No.	Answer	Detailed Explanation
1	C	$3^2+4^2=5^2$ $9+16=25$ $25=25$
2	B	$8^2+15^2=c^2$ $64+225=c^2$ $289=c^2$ 17 cm $= c$
3	C	$7^2+11^2=c^2$ $49+121=c^2$ $170=c^2$ $13.0 \approx c$
4	C	$5^2+5^2=c^2$ $25+25=c^2$ $50=c^2$ 7 inches $\approx c$
5	D	$a^2+b^2=c^2$ cannot be changed to $a^2+c^2=b^2$.
6	C	$9^2+12^2=15^2$ $81+144=225$ $225=225$
7	A	$4^2+4^2= c^2$ $32 = c^2$ 5.7 in $\approx c$
8	B	$16+a^2 = 100$ $a^2 = 84$ $a \approx 9.2$
9	C	$AC^2+BC^2=AB^2$ $AC^2+28^2=35^2$ $AC^2=1225-784$ $AC^2=441$ $AC=21$
10	D	$2s^2=625$ $s^2=312.5$ s is approximately 18

Question No.	Answer	Detailed Explanation
11	B	The equation will be $6^2 + 8^2 = c^2$. When you square both of those you get $100 = c^2$. To isolate c, you have to take the square root of both sides. When you do this you have $\sqrt{100} = C$. So once you take the square root, you are left with $c = 10$.
12	B	In order to solve this equation, you need to have 15 as the hypotenuse. This means that it is one the opposite side of the equal sign. The equation you should select is $4^2 + x^2 = 15^2$.

Lesson 9: Pythagorean Theorem in Real-World Problems

Question No.	Answer	Detailed Explanation
1	C	$8^2+h^2=17^2$ $h^2=289-64$ $h^2=225$ h=15 feet
2	A	Applying the Pythagorean Theorem to this (w, 18, 20) right triangle, $w^2 + 18^2 = 20^2$ is the correct equation.
3	D	$5^2+3^2=$(diagonal of bottom of chest)^2 $\sqrt{34}$=diagonal of bottom of chest $34+9=d^2$ $43=d^2$ $\sqrt{43}$ ft =d
4	A	Remember the Pythagorean triples (3,4,5 and their multiples). In this case: 30, 40, 50
5	B	$9^2+20^2=c^2$ $81+400=c^2$ $481=c^2$ $\sqrt{481}=c$ c is approximately 22 ft
6	B	$a^2+a^2=50^2$ $2a^2=2500$ $a^2=1250$ a=35 yd (approx)
7	A	$30^2+20^2=d^2$ $1300=d^2$ $\sqrt{1300}=d$ d=approx. 36 ft
8	B	To find the shortcut distance, think of the two walking paths (20 ft and 15 ft) as the legs of a right triangle. The straight-line path is the hypotenuse: $20^2 + 15^2 = 400 + 225 = 625$ and square root of 625 = 25 ft The original walking distance was: 20 + 15 = 35 ft So the shortcut saves: 35 − 25 = 10 ft
9	C	Use the Pythagorean triple: 30, 40, 50 He can place his ladder up to 30 feet away.

<table>
<tr><th>Question No.</th><th>Answer</th><th>Detailed Explanation</th></tr>
<tr><td>10</td><td>D</td><td>$25^2+x^2=50^2$
$x^2=2500-625=1875$
x=43 ft (approx)
Alternate Method :
$x^2 = 50^2 - 25^2$
$x^2 = 25^2 \times 2^2 - 25^2$
$x^2 = 25^2 (2^2 - 1)$
$x = \sqrt{(25^2 \times 3)}$
$x = 25 \times \sqrt{3}$
$x = 25 \times 1.73$ (when $\sqrt{3}$ is rounded to the nearest hundredth)
x = 43 ft (approx)</td></tr>
<tr><td>11</td><td></td><td>
<table>
<tr><th></th><th>$9^2+x^2=15^2$</th><th>$40^2+38^2=x^2$</th><th>$9^2+15^2=x^2$</th><th>$x^2+38^2=40^2$</th></tr>
<tr><td>One house is 15 miles due north of the park. Another house is 9 miles due east of the park. How far apart are the houses from each other?</td><td></td><td></td><td>●</td><td></td></tr>
<tr><td>The foot of a ladder is put 9 feet from the wall. If the ladder is 15 feet long how high up the building will the ladder reach?</td><td>●</td><td></td><td></td><td></td></tr>
<tr><td>If you drive your car 40 miles south and then 38 miles east, how far would the shortest route be from your starting point?</td><td></td><td>●</td><td></td><td></td></tr>
<tr><td>The diagonal of a TV is 40 inches. The TV is 38 inches long. How tall is the TV?</td><td></td><td></td><td></td><td>●</td></tr>
</table>
The first problem provides the two legs of the triangle, so you need to use the equation to solve for the hypotenuse. In the second problem, the ladder is the hypotenuse, so you will solve for one of the legs. The third problem gives you the legs, so you solve for the hypotenuse. In the last problem, the diagonal is the hypotenuse, so you solve for a leg.</td></tr>
</table>

Question No.	Answer	Detailed Explanation
12	(1) 10, 26 (2) 30 (3) 5, 11.2	(1) The first example you are looking for the other leg. Since the base length is 48, you have to cut it in half to make the right triangle. So in this case, one of the legs will be 24 cm. The hypotenuse is going to be the slant height of 26 cm. This means you are solving for the other leg. When you fill the numbers into Pythagorean Theorem you get 10 cm. (2) The next example you have both a leg and a hypotenuse, so you are solving for the other leg. In this case the other leg will be 15 in. which is half of the base length. So, base length will be 30 in. (3) The last example you have both legs, 5 ft. and 10 ft. So the length of the ladder will be the hypotenuse. In this case it will be about 11.2 ft.
13	C	The 2 directions you biked formed the legs of the triangle. So you are solving for the hypotenuse. To solve this, square the lengths of the two legs and add them together. The last step is to take the square root of the sum. In this case, third choice is the correct answer.

Lesson 10: Pythagorean Theorem & Coordinate System

Question No.	Answer	Detailed Explanation
1	A	$4^2 + 7^2 = d^2$ $65 = d^2$; d is approx. 8 m
2	C	Draw the graph. Draw the right triangle with the distance between the points as the hypotenuse. The legs will be 6 units & 8 units. Find the hypotenuse. $8^2 + 6^2 = h^2$ $64 + 36 = h^2$ $100 = h^2$; $10 = h$
3	B	$4^2 + 14^2 = d^2$ $16 + 196 = d^2$ $212 = d^2$ d = 14.6 (approximately)
4	C	$25 + 9 = 34 = d^2$ d = 5.8 (approximately)
5	C	$4^2 + 4^2 = d^2$ $32 = d^2$ d = 5.7 (approximately)
6	B	$2^2 + 10^2 = AB^2$ $\sqrt{104} = AB$ 10.2 = AB (approximately) $4^2 + 9^2 = AC^2$ $\sqrt{97} = AC$ 9.8 = AC (approximately) Note that, we can conclude AB > AC without calculating square root also, because $AB^2 > AC^2$ implies AB > AC
7	A	$50^2 + 40^2 = d^2$ $2500 + 1600 = d^2$ $4100 = d^2$ 64 yd is approx. distance
8	B	Original point: (7, 8) New point: (8, -7) Final point: (-8, -7) $15^2 + 15^2 = d^2$ $450 = d^2$ d = 21.2 units (approximately)

<table>
<tr><th>Question No.</th><th>Answer</th><th>Detailed Explanation</th></tr>
<tr><td>9</td><td>B</td><td>Draw a sketch. The coordinates of the rectangle are (0,0), (15,0), (15,20), and (0,20). The midpoint of the width is 7.5 and the midpoint of the length is 10. Therefore, the center of the rectangle is at (7.5, 10).</td></tr>
<tr><td>10</td><td>B</td><td>$3^2 + 3^2 = d^2$
$18 = d^2$; $4.2 = d$ (approximately)</td></tr>
<tr><td>11</td><td></td><td>
<table>
<tr><th></th><th>10.8</th><th>12.2</th><th>13</th><th>14.8</th></tr>
<tr><td>(6, 5) and (-4, 9)</td><td>●</td><td></td><td></td><td></td></tr>
<tr><td>(-8, 0) and (5, -7)</td><td></td><td></td><td></td><td>●</td></tr>
<tr><td>(-4, -9) and (6, -2)</td><td></td><td>●</td><td></td><td></td></tr>
<tr><td>(5, 4) and (12, 15)</td><td></td><td></td><td>●</td><td></td></tr>
</table>

Distance between two points having coordinates (x_1,y_1) and (x_2,y_2) can be found out with the help of Pythagorean Theorem. $D=\sqrt{(x_2-x_1)^2+(y_2-y_1)^2}$. Using this formula we get -
(i) d = 10.8
(ii) d = 14.8
(iii) d = 12.2
(iv) d = 13</td></tr>
<tr><td>12</td><td>8||10||1.41||1||4</td><td>The first set of ordered pairs have the missing leg length of 8 which we can then fill those lengths into Pythagorean Theorem. So the hypotenuse is 10. The second set you already were given both leg lengths so they need to be filled into Pythagorean Theorem. This will make the hypotenuse 1.41 (approximately). The last set already gives you the hypotenuse so you just need to plot the ordered pairs and form your triangle. When you do this, you will see that the legs lengths are 1 and 4.</td></tr>
</table>

Lesson 11: Finding Volume: Cone, Cylinder, & Sphere

Question No.	Answer	Detailed Explanation
1	D	$V=\frac{4}{3}(\pi r^3)$ $V=\frac{4}{3}(\pi)\ 6^3$ $V=\frac{4}{3}(216)\ \pi$ $V=288\pi$
2	C	$V=(\frac{1}{3})Bh$ where B= area of the base $V=(\frac{1}{3})\pi(4^2)(9)$ $V=48\pi$
3	C	$V=\pi r^2h$ $V=\pi\ (25)(3)$ $V=75\pi$
4	D	$V=(\frac{4}{3})\pi r^3$ $V=\frac{4}{3}(125\pi)$ $V=(\frac{500}{3})\pi$ $V=523.6$ cubic inches
5	B	$V=\pi r^2h$ $V=(5)(2)^2\pi$ $V=20\pi$ $V=20(3.14)$ cu in $V=62.8$ cu in = 60 cu in (when rounded to the nearest ten)
6	D	$V_{cylinder} = \pi r^2h$ $V_{cone}=\frac{1}{3}\ \pi r^2h$ If volumes are the same, then the height of the cone must be 3 times the height of the cylinder because $3(\frac{1}{3})=1$.

Question No.	Answer	Detailed Explanation
7	C	$V_{sphere}=\frac{4}{3}\pi r^3$ $V=\frac{4}{3}(8)\pi$ $V=\frac{32}{3}\pi$ $V_{cylinder}=\pi r^2h$ $V=8\pi$ $V_{cone}=\frac{1}{3}\pi r^2h$ $V=(\frac{1}{3})(9)(4)\pi$ $V=12\pi$; The cone has the greatest volume.
8	D	$V_{sphere}=\frac{4}{3}\pi r^3$ $V=\frac{4}{3}\pi(27)$ $V=36\pi$ $V_{cylinder}=\pi r^2h$ $V=9(4)\pi$ $V=36\pi$ $V_{cone}=\frac{1}{3}\pi r^2h$ $V=(\frac{1}{3})(36)(3)\pi$ $V=36\pi$
9	A	$V_{sphere}/V_{cone} = (\frac{4}{3})\pi r^3 / (\frac{1}{3})\pi r^2 h = 4r/h = 4r/4 = r/1$
10	A	$V_{cylinder}/V_{cone}= \pi r^2h/(\frac{1}{3})\pi r^2h=3/1$
11	75.36 ft^3	Radius of the cone = r = 3 ft. Height of the cone = h = 8 ft. Volume of the cone = $V= (1/3)\pi r^2h$ $V = (1/3) \times 3.14 \times 3^2 \times 8$ $V = 75.36\ ft^3$
12	B	The formula used to find the volume of a sphere is $V= (\frac{4}{3})\pi r^3$. You just have to substitute the radius of 7 into the equation for r. Thus making it the second choice.

Date of Completion:______________ Score:______________

Chapter 5: Statistics and Probability

Lesson 1: Interpreting Data Tables & Scatter Plots

1. If a scatter plot has a line of best fit that decreases from left to right, which of the following terms describes the association?

Ⓐ Positive association
Ⓑ Negative association
Ⓒ Constant association
Ⓓ Nonlinear association

2. If a scatter plot has a line of best fit that increases from left to right, which of the following terms describes the association?

Ⓐ Positive association
Ⓑ Negative association
Ⓒ Constant association
Ⓓ Nonlinear association

3. Which of the following scatter plots is the best example of a linear association?

Ⓐ
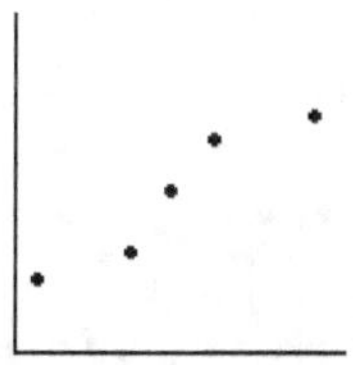

Ⓑ
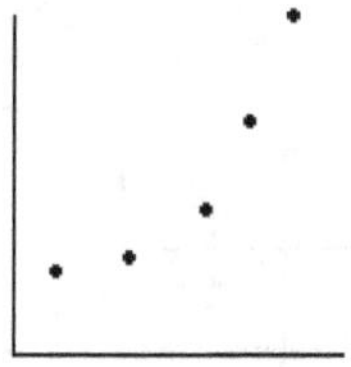

Ⓒ
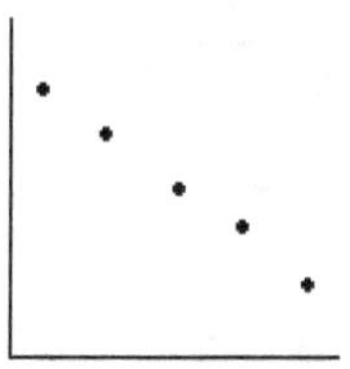

Ⓓ
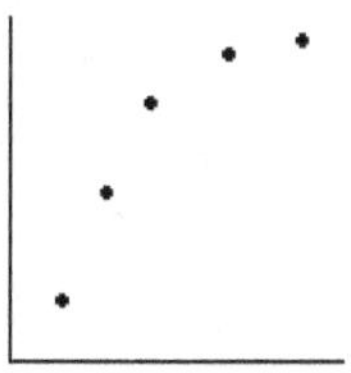

4. **Data for 9 kids' History and English grades are made available in the chart. What is the association between the History and English grades?**

Kids	1	2	3	4	5	6	7	8	9
History	63	49	84	33	55	23	71	62	41
English	67	69	82	32	59	26	73	62	39

Ⓐ Positive association
Ⓑ Negative association
Ⓒ Nonlinear association
Ⓓ Constant association

5. **Data for 9 kids' History grades and the distance they live from school are made available in the chart. What is the association between these two categories?**

Kids	1	2	3	4	5	6	7	8	9
History	63	49	84	33	55	23	71	62	41
Distance from School (miles)	.5	7	3	4	5	2	3	6	9

Ⓐ No association
Ⓑ Positive association
Ⓒ Negative association
Ⓓ Constant association

6. **Data for 9 kids' Math and Science grades are made available in the chart. What is the association between the Math and Science grades?**

Kids	1	2	3	4	5	6	7	8	9
Science	63	49	84	33	55	23	71	62	41
Math	67	69	82	32	59	26	73	62	39

Ⓐ Positive association
Ⓑ No association
Ⓒ Constant association
Ⓓ Negative association

7. Which of the scatter plots below is the best example of positive association?

Ⓐ 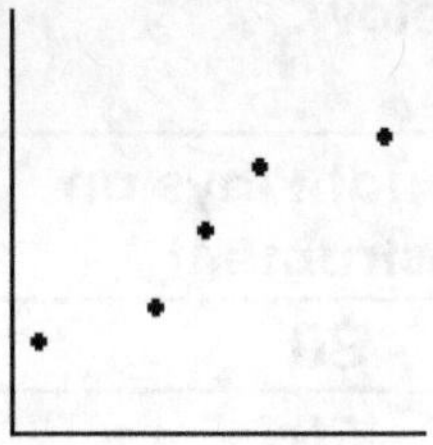

Ⓑ 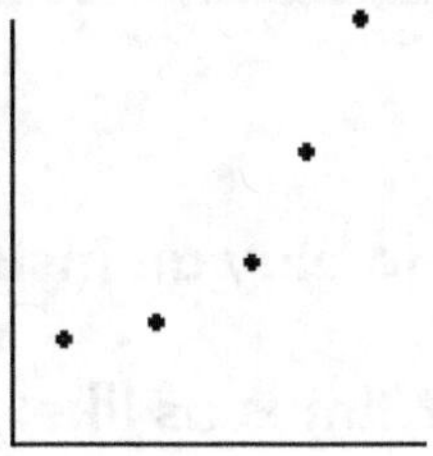

Ⓒ 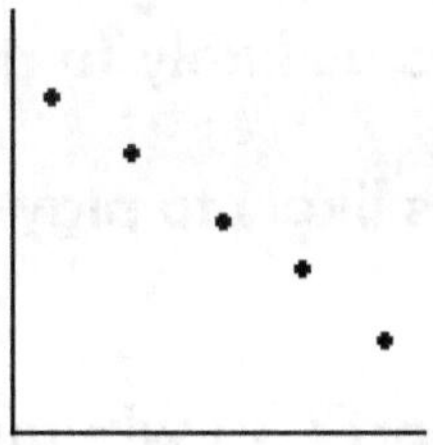

Ⓓ 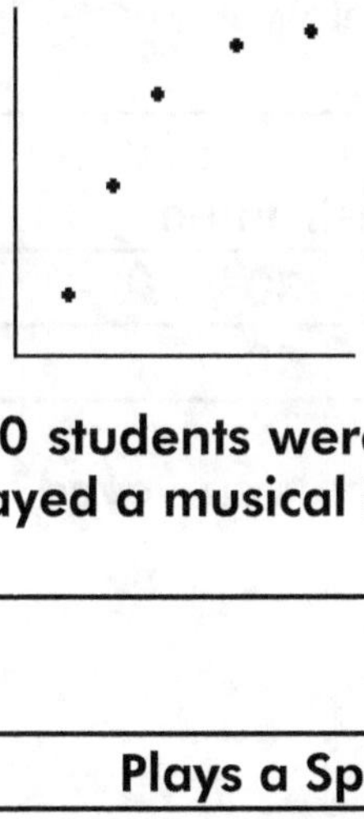

8. 150 students were surveyed and asked whether they played a sport and whether they played a musical instrument. The results are shown in the table below.

	Plays an Instrument	Does not Play an Instrument
Plays a Sport	60	30
Does not Play a Sport	10	50

What percent of the 150 students play a sport and also do not play an instrument?

Ⓐ 20%
Ⓑ 33%
Ⓒ 40%
Ⓓ 50%

9. 150 students were surveyed and asked whether they played a sport and whether they played a musical instrument. The results are shown in the table below.

	Plays an Instrument	Does not Plays an Instrument
Plays a Sport	60	30
Does not Play a Sport	10	50

Which of the following statements is NOT supported by the data?

Ⓐ A randomly chosen student who plays a sport is 2 times as likely to play an instrument as to not play an instrument.
Ⓑ A randomly chosen student who does not play an instrument is 2 times as likely to not play a sport as to play a sport.
Ⓒ A randomly chosen student who does not play a sport is 5 times as likely to not play an instrument as to play an instrument.
Ⓓ A randomly chosen student who plays an instrument is 6 times as likely to play a sport as to not play a sport.

10. 150 students were surveyed and asked whether they played a sport and whether they played a musical instrument. The results are shown in the table below.

	Plays an Instrument	Does not Plays an Instrument
Plays a Sport	60	30
Does not Play a Sport	10	50

Which two sections add up to just over half of the number of students surveyed?

Ⓐ The two sections that do not play an instrument.
Ⓑ The two sections that do not play a sport.
Ⓒ The two sections that play an instrument.
Ⓓ The two sections that play a sport.

11. Match the data with the correct association.

<table>
<tr><th></th><th>POSITIVE ASSOCIATION</th><th>NEGATIVE ASSOCIATION</th><th>NO ASSOCIATION</th></tr>
<tr><td>The population survey data for 5 years shows the number of goldfish and star fish. Describe the association between the population of goldfish and star fish.
<table>
<tr><td>YEAR</td><td>1</td><td>2</td><td>3</td><td>4</td><td>5</td></tr>
<tr><td>GOLDFISH</td><td>13</td><td>18</td><td>19</td><td>20</td><td>25</td></tr>
<tr><td>STARFISH</td><td>30</td><td>25</td><td>20</td><td>15</td><td>12</td></tr>
</table></td><td>○</td><td>○</td><td>○</td></tr>
<tr><td>Below is data for 5 years showing Jonny's and Jack's weight in kg. Describe the association between the weight of Jonny and Jack.
<table>
<tr><td>YEAR</td><td>1</td><td>2</td><td>3</td><td>4</td><td>5</td></tr>
<tr><td>JONNY</td><td>30</td><td>40</td><td>42</td><td>49</td><td>52</td></tr>
<tr><td>JACK</td><td>30</td><td>35</td><td>40</td><td>45</td><td>50</td></tr>
</table></td><td>○</td><td>○</td><td>○</td></tr>
<tr><td>The data for 5 days shows the sale of watermelon and potatoes. Describe the association between the sale of watermelon and potatoes.
<table>
<tr><td>DAYS</td><td>1</td><td>2</td><td>3</td><td>4</td><td>5</td></tr>
<tr><td>WATERMELON</td><td>70</td><td>26</td><td>60</td><td>19</td><td>70</td></tr>
<tr><td>POTATO</td><td>10</td><td>40</td><td>15</td><td>80</td><td>22</td></tr>
</table></td><td>○</td><td>○</td><td>○</td></tr>
</table>

12. Following is 10 days of data which shows the sale of apples and mangoes. Fill in the type of association there is between the apple and mango sales.

DAYS	1	2	3	4	5	6	7	8	9	10
APPLE	62	49	81	26	45	55	16	74	97	34
MANGO	36	44	49	37	26	11	76	83	64	81

There is ________________________ between apple and mango sale.

13. **Fill in the table with the word positive, negative, or none to describe which type of association is plotted.**

Scatter Plot	Type of Association
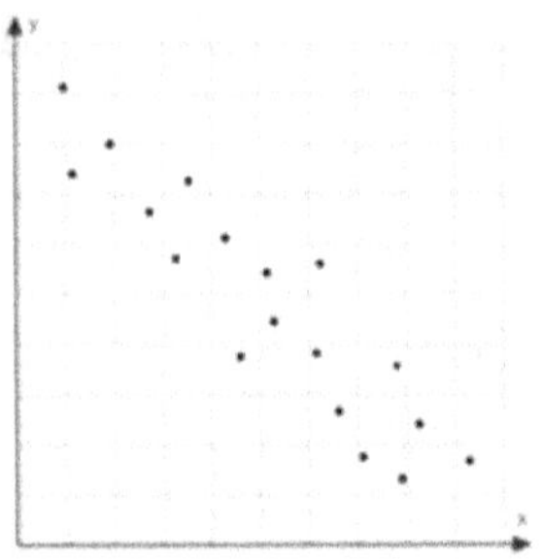	
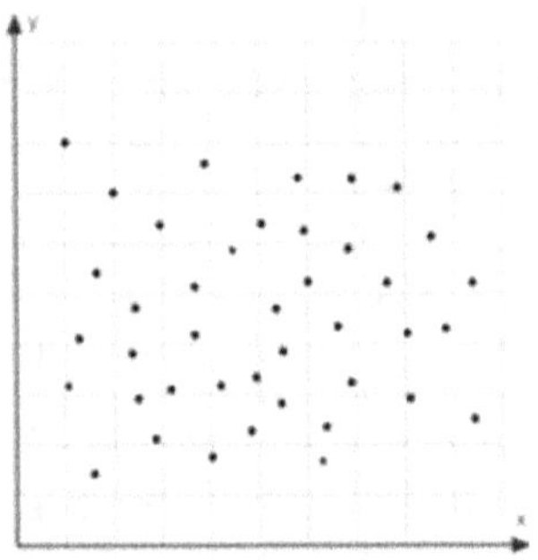	
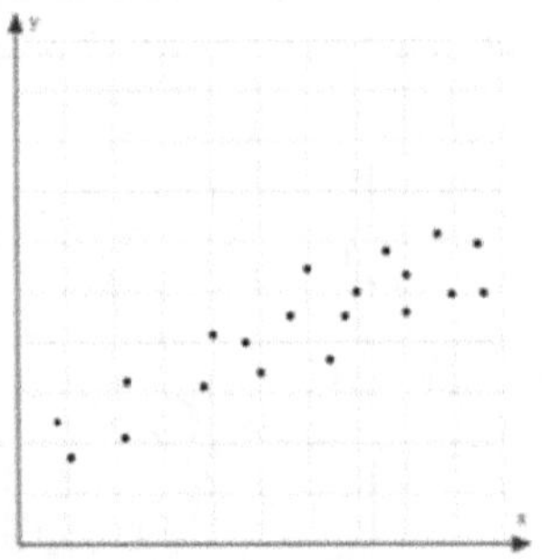	
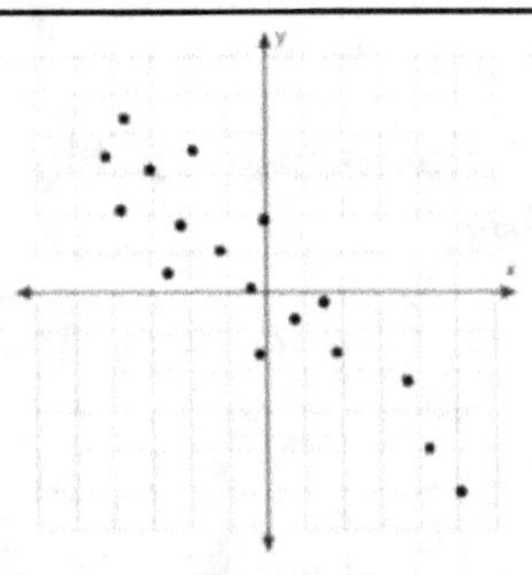	

CHAPTER 5 → Lesson 2: Scatter Plots, Line of Best Fit

1. 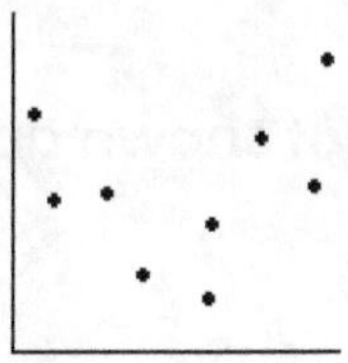

Which of the following best describes the points in this scatter plot?

Ⓐ Increasing Linear
Ⓑ Decreasing Linear
Ⓒ Constant Linear
Ⓓ None of these

2. 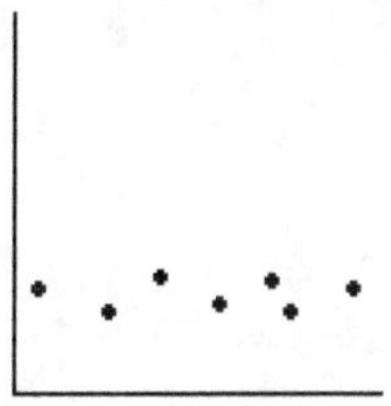

Which of the following best describes the points in this scatter plot?

Ⓐ Increasing Linear
Ⓑ Decreasing Linear
Ⓒ Constant Linear
Ⓓ None of these

3. 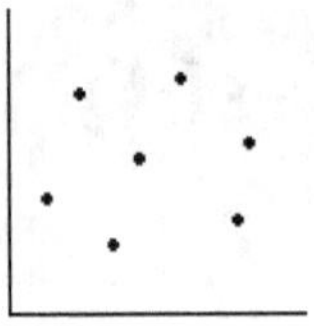

Which of the following lines best approximates the data in the scatter plot shown above?

Ⓐ 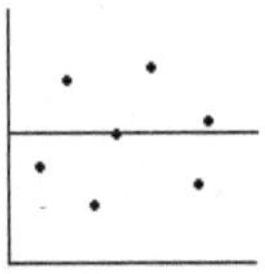

Ⓑ 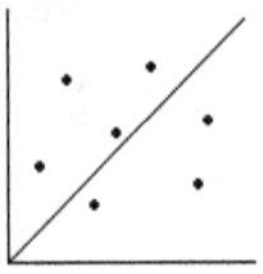

Ⓒ 

Ⓓ **None of these; the data do not appear to be related linearly.**

4. **Which scatter plot represents a positive linear association?**

Ⓐ 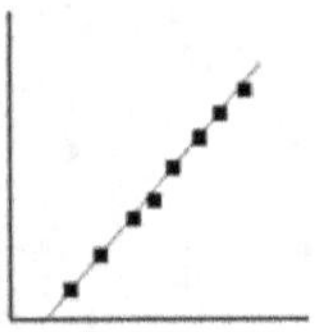

Ⓑ 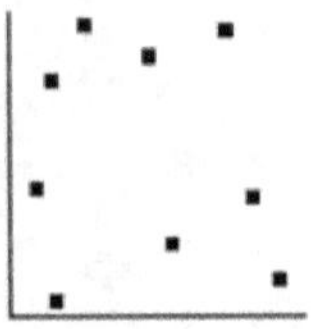

Ⓒ 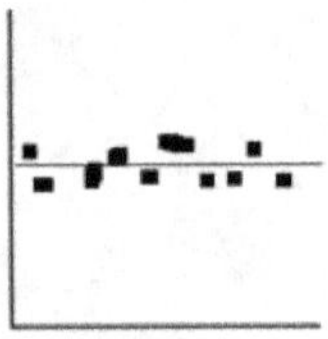

Ⓓ 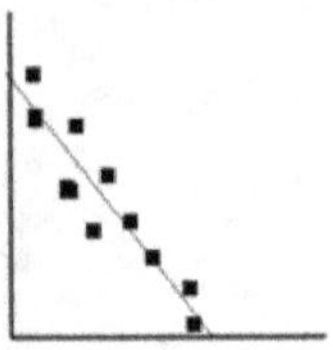

5. **Which scatter plot represents a negative linear association?**

Ⓐ 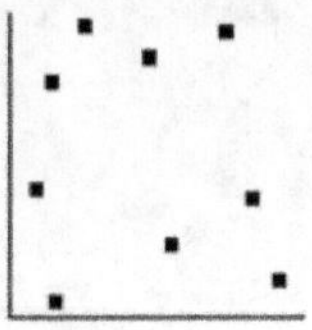

Ⓑ 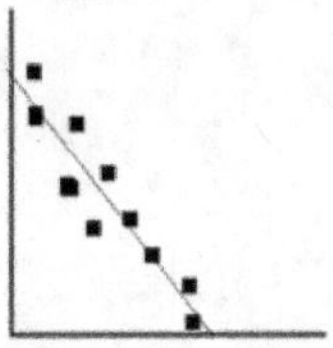

Ⓒ 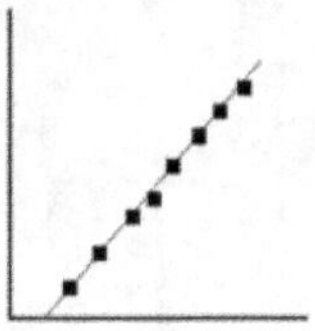

Ⓓ 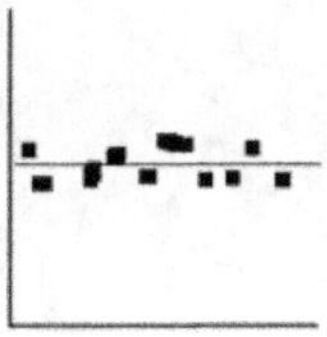

6. **Which scatter plot represents no association?**

Ⓐ 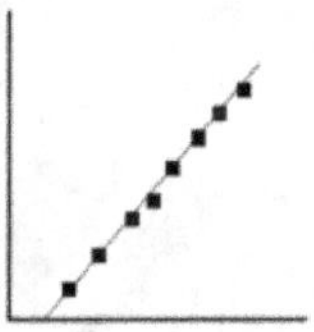

Ⓑ 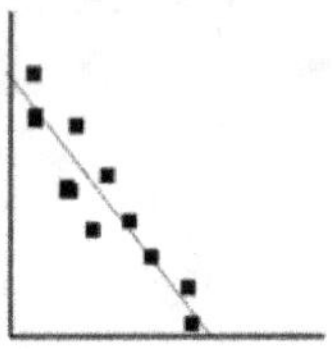

Ⓒ 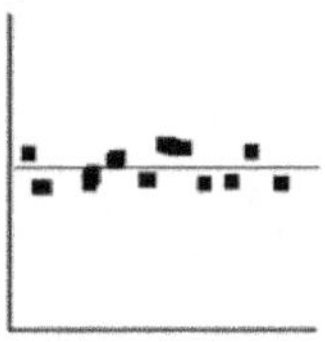

Ⓓ 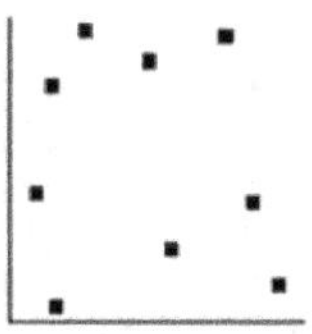

7. **Which scatter plot represents a constant association?**

Ⓐ 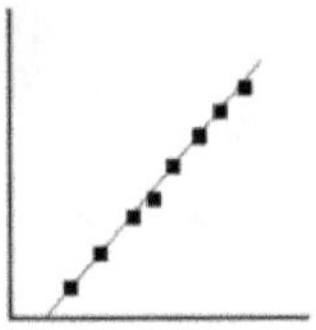

Ⓑ 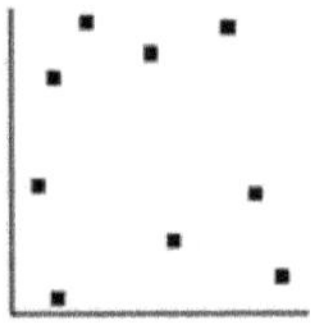

Ⓒ 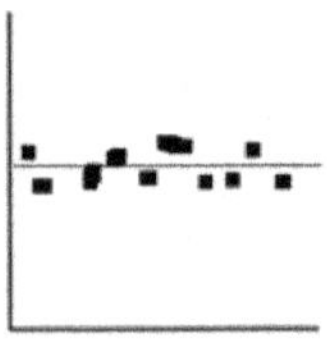

Ⓓ 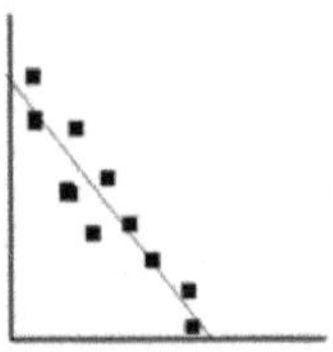

8. **The graph of this data set would most resemble which of the following graphs?**

x	1	2	3	4	5	6	7
y	2	3	4	5	6	7	8

Ⓐ 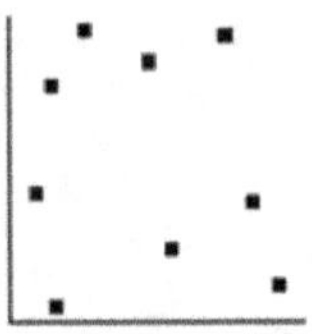

Ⓑ 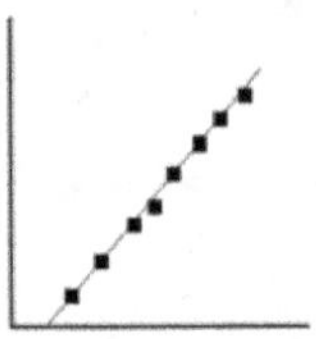

Ⓒ 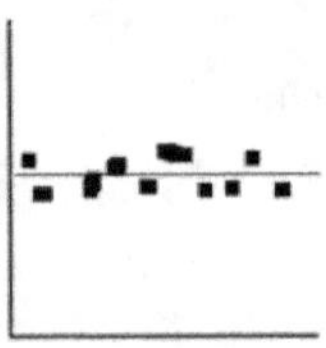

Ⓓ 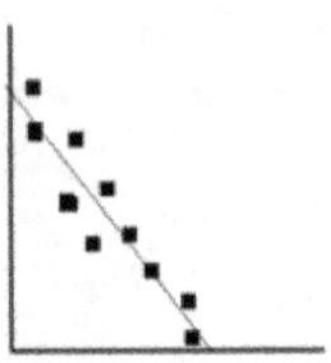

9. **The graph of this data would most resemble which of the following graphs?**

x	1	2	3	4	5	6	7
y	4	4	4	4	4	4	4

Ⓐ 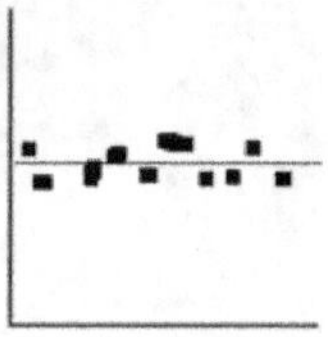

Ⓑ 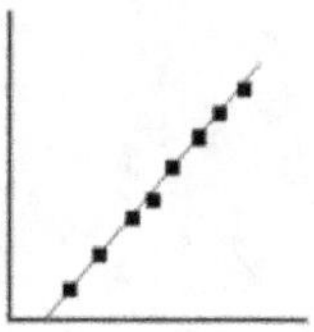

Ⓒ 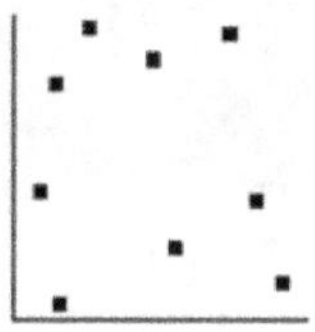

Ⓓ 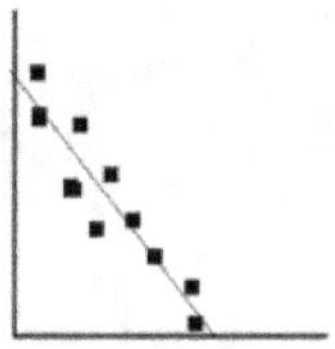

10. **Typically air temperature decreases through the night between midnight and 6:00 am. This is an example of what type of association?**

Ⓐ constant association
Ⓑ positive linear association
Ⓒ negative linear association
Ⓓ no association

11. Match which line would be the best fit to describe the data pictured.

Figure - 1

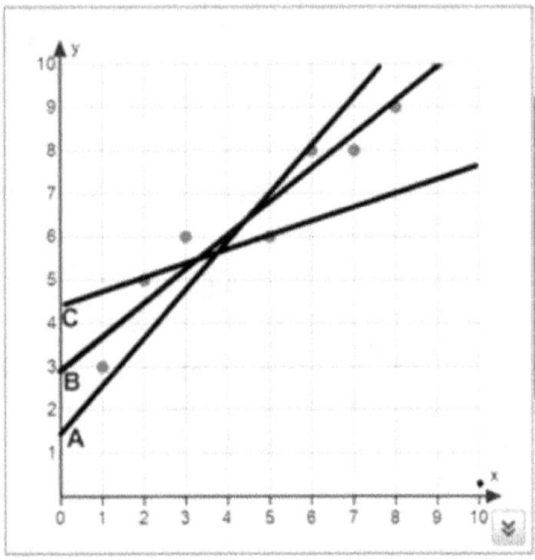

X	Y
1	3
2	5
3	6
5	6
6	8
7	8
8	9
9	10
10	11

Figure - 2

X	Y
1	9
2	8
3	6
5	6
6	5
7	4
8	4
9	2
10	0

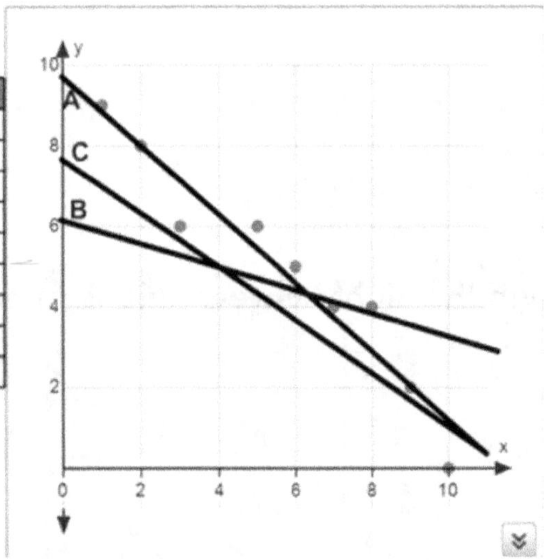

	A	B	C
Figure - 1	○	○	○
Figure - 2	○	○	○

12. Write the equation of the line that best fits the graph using the two labeled points. Leave your answer as fractions.

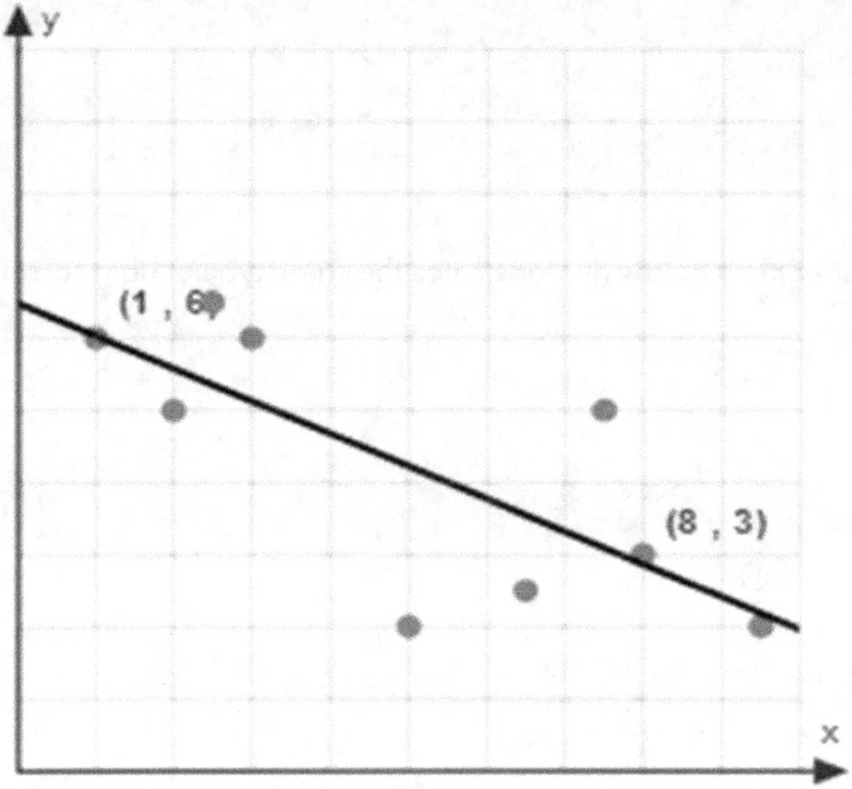

y= [ ] x + [ ]

13. Cathy wanted to know what kind of shows the 8th grade class preferred – dramas or comedies. 55 students said they liked comedies and not dramas. 25 students liked both dramas and comedies. There were 41 students who did not like dramas nor comedies. Complete a two way table using the information given.

	Doesn't Like Dramas	Likes Dramas	Total
Doesn't Like Comedies	41		97
Likes Comedies		25	
Total	96		177

CHAPTER 5 → Lesson 3: Analyzing Linear Scatter Plots

1. 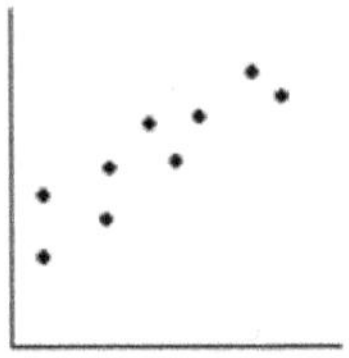

Which of the following scatter plots below demonstrates the same type of data correlation as the one shown above?

Ⓐ 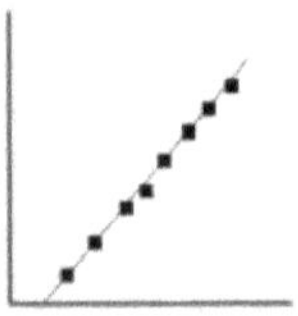

Ⓑ 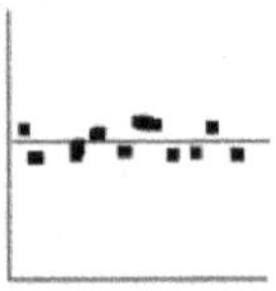

Ⓒ 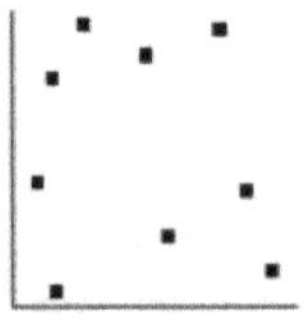

Ⓓ 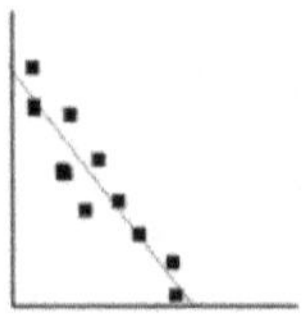

2. 

Which of the following lines most accurately models the points in this scatter plot?

Ⓐ 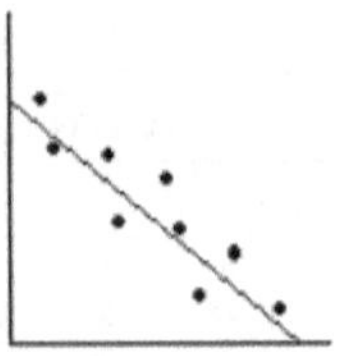

Ⓑ 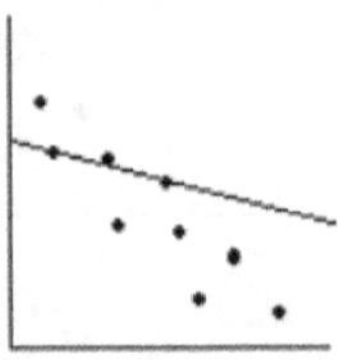

Ⓒ 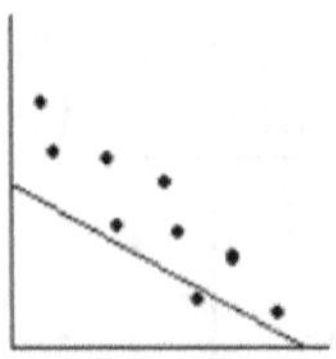

Ⓓ 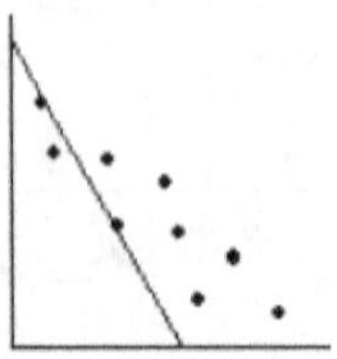

3. 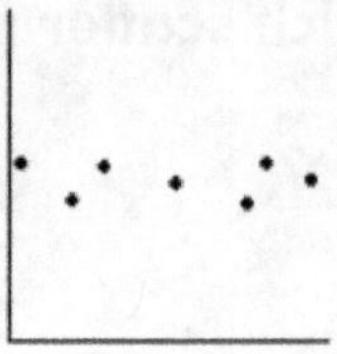

Which of the following lines most accurately models the points in this scatter plot?

Ⓐ 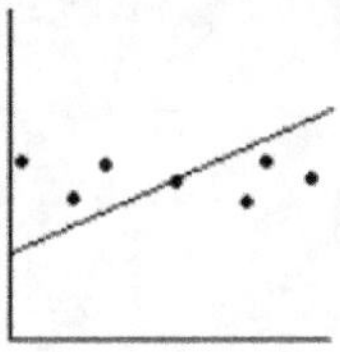

Ⓒ 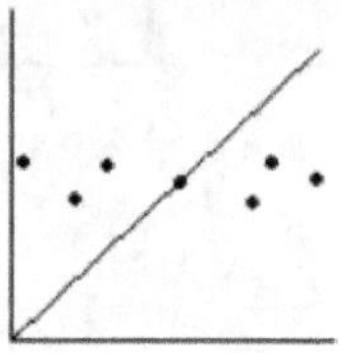

Ⓑ 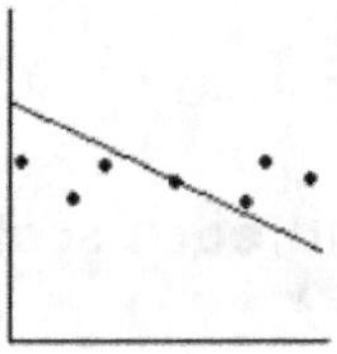

Ⓓ 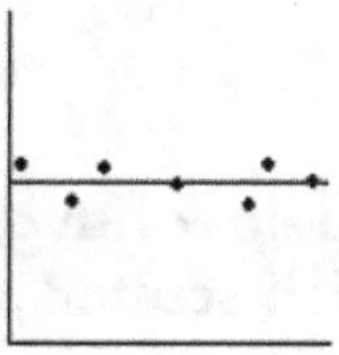

4. 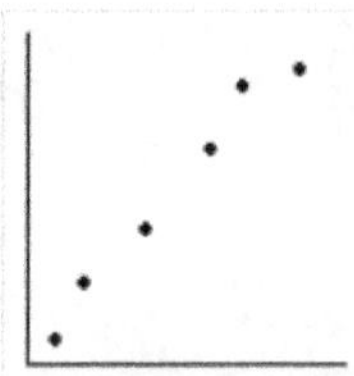

Which of the following lines most accurately models the points in this scatter plot?

Ⓐ 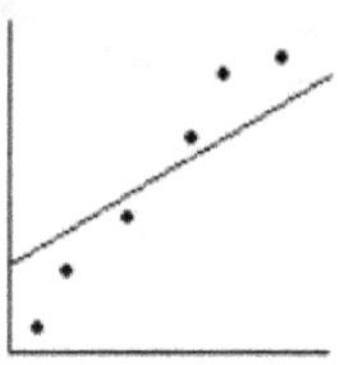

Ⓒ 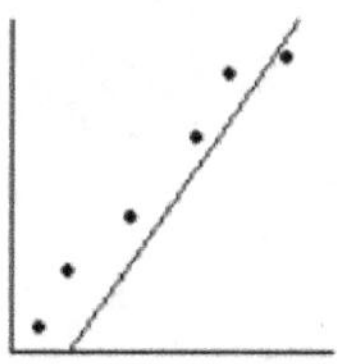

Ⓑ 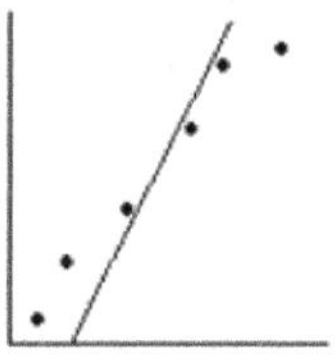

Ⓓ 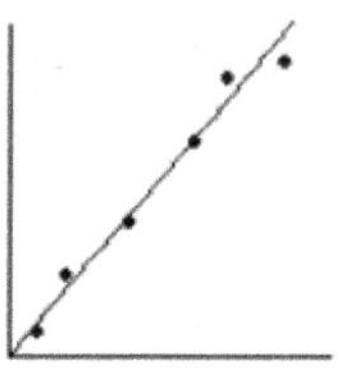

5. **The four scatter plots below have all been modeled by the same line. Which scatter plot has the strongest association?**

Ⓐ 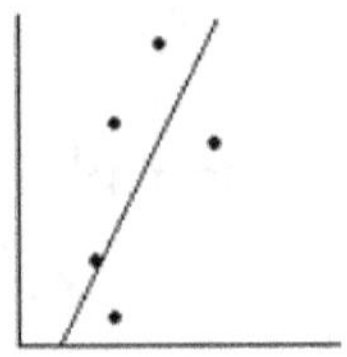

Ⓑ 

Ⓒ 

Ⓓ 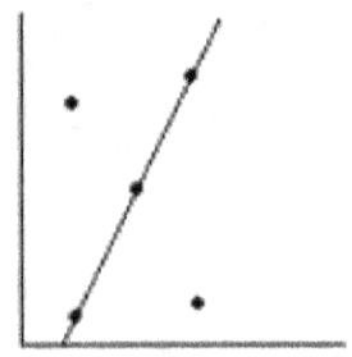

6. **The four scatter plots shown below have four points in common, and each scatter plot has a different fifth point. Which scatter plot's fifth point is an outlier?**

Ⓐ 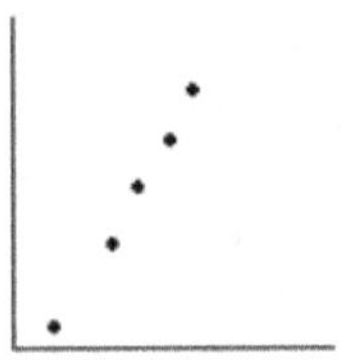

Ⓑ 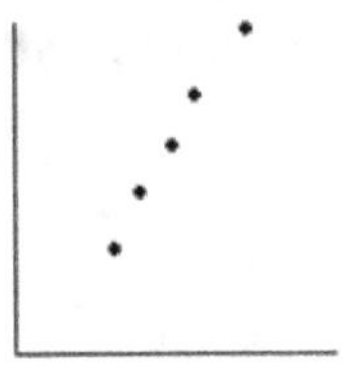

Ⓒ 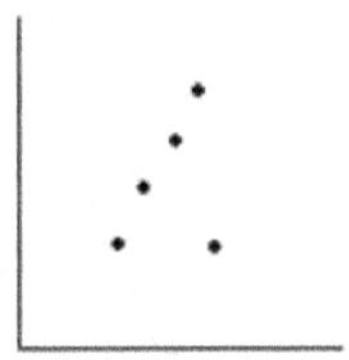

Ⓓ 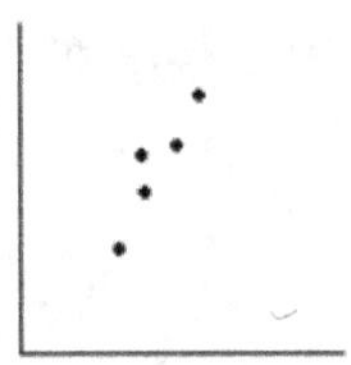

7. **The four scatter plots shown below have four points in common, and each scatter plot has a different fifth point. Which scatter plot's fifth point is NOT an outlier?**

Ⓐ 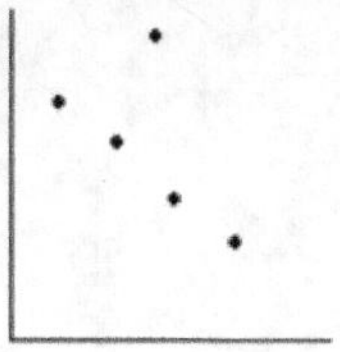

Ⓒ 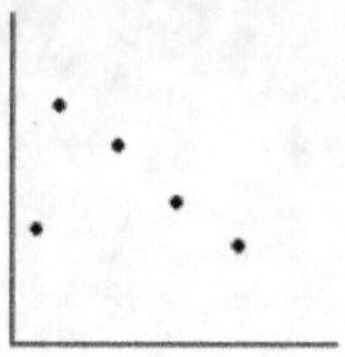

Ⓑ 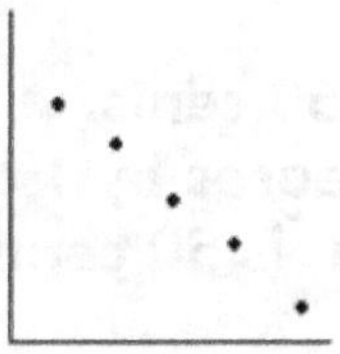

Ⓓ 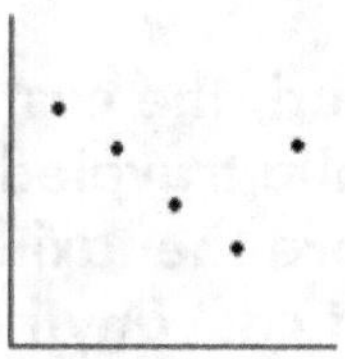

8. **The figure below shows a scatter plot relating the length of a bean plant, in centimeters, to the number of days since it was planted. The slope of the associated line is 2. Which of the following correctly interprets the slope?**

Ⓐ **The bean plant grows approximately 1 cm every 2 days.**
Ⓑ **The bean plant grows approximately 2 cm each day.**
Ⓒ **The bean plant was 2 cm long when it was planted.**
Ⓓ **The bean plant approximately doubles in length each day.**

9. The figure below shows a scatter plot relating the cost of a ride in a taxicab, in dollars, to the number of miles traveled. The slope of the associated line is 0.5. Which of the following correctly interprets the slope?

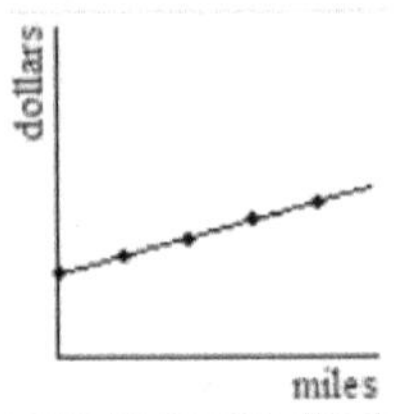

Ⓐ For each additional mile traveled, the cost of the ride increases by 50 cents.
Ⓑ For each additional half of a mile traveled, the cost of the ride increases by 1 dollar.
Ⓒ The initial cost of the ride, before the taxi has traveled any distance, is 50 cents.
Ⓓ The first half of a mile does not cost anything.

10. The figure below shows a scatter plot relating the temperature in a school's parking lot, in degrees Fahrenheit, to the number of hours past noon. The slope of the associated line is -3. Which of the following correctly interprets the slope?

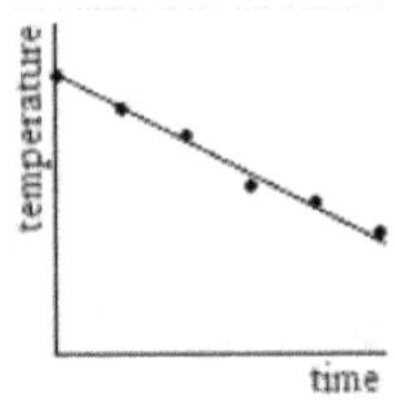

Ⓐ The temperature at noon was -3 degrees Fahrenheit.
Ⓑ The temperature decreased until it reached -3 degrees Fahrenheit.
Ⓒ The temperature decreased an average of 1 degree Fahrenheit every 3 hours.
Ⓓ The temperature decreased an average of 3 degrees Fahrenheit per hour.

11. Match the correct vocab term with the correct definition.

	Linear	Negative Association	Line of Best Fit	Prediction Equation
A line on a graph showing the general direction that a group of points seem to be heading	○	○	○	○
A graph that is represented by a straight line	○	○	○	○
The equation of a line that can predict outcomes using given data	○	○	○	○
A correlation of points that is linear with a negative slope	○	○	○	○

12. Write the equation of the best fit line for this scatter plot using the 2 ordered pairs given.

Write your answer in the box given below.

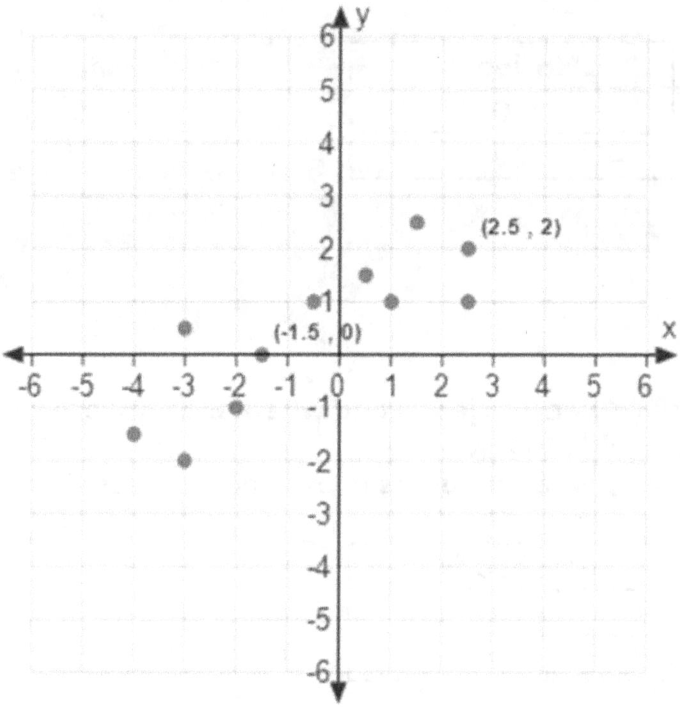

CHAPTER 5 → Lesson 4: Relatable Data Frequency

1. 50 people were asked whether they were wearing jeans and whether they were wearing sneakers. The results are shown in the table below.
What percent of the people who wore sneakers were also wearing jeans?

	Jeans	No Jeans
Sneakers	15	10
No Sneakers	5	20

Ⓐ 15%
Ⓑ 30%
Ⓒ 60%
Ⓓ 75%

2. 50 people were asked whether they were wearing jeans and whether they were wearing sneakers. The results are shown in the table below.
What percent of the people who wore jeans were also wearing sneakers?

	Jeans	No Jeans
Sneakers	15	10
No Sneakers	5	20

Ⓐ 15%
Ⓑ 30%
Ⓒ 60%
Ⓓ 75%

3. 50 people were asked whether they were wearing jeans and whether they were wearing sneakers. The results are shown in the table below.
What percent of the people who did NOT wear sneakers were wearing jeans?

	Jeans	No Jeans
Sneakers	15	10
No Sneakers	5	20

Ⓐ 5%
Ⓑ 20%
Ⓒ 25%
Ⓓ 40%

4.

	Fertilizer	No Fertilizer
Lived	200	600
Died	50	150

Out of 1,000 plants, some were given a new fertilizer and the rest were given no fertilizer. Some of the plants lived and some of them died, as shown in the table above. Which of the following statements is supported by the data?

Ⓐ Fertilized plants died at a higher rate than unfertilized plants did.
Ⓑ Fertilized plants and unfertilized plants died at the same rate.
Ⓒ Fertilized plants died at a lower rate than unfertilized plants died.
Ⓓ None of the above statements can be supported by the data.

5.

	Windy	Not Windy
Sunny	5	15
Cloudy	4	6

The weather was observed for 30 days; each day was classified as sunny or cloudy, and also classified as windy or not windy. The results are shown in the table above. Which of the following statements is NOT supported by the data?

Ⓐ 25% of the sunny days were also windy.
Ⓑ 30% of the days were windy.
Ⓒ 40% of the cloudy days were also windy.
Ⓓ 50% of the windy days were also sunny.

6.

	Jeans	No Jeans
Sneakers	15	10
No Sneakers	5	20

50 people were asked whether they were wearing jeans and whether they were wearing sneakers. The results are shown in the table above.
What fraction of the people who wore sneakers were NOT wearing jeans?

Ⓐ $\frac{1}{5}$
Ⓑ $\frac{2}{5}$
Ⓒ $\frac{3}{10}$
Ⓓ $\frac{3}{4}$

7.

	Jeans	No Jeans
Sneakers	15	10
No Sneakers	5	20

50 people were asked whether they were wearing jeans and whether they were wearing sneakers. The results are shown in the table above. Which of the following statements is NOT supported by the data?

Ⓐ A randomly chosen person who is wearing sneakers is equally as likely to be wearing jeans as not wearing jeans.
Ⓑ A randomly chosen person who is not wearing jeans is 2 times as likely to be not wearing sneakers as wearing sneakers.
Ⓒ A randomly chosen person who is wearing jeans is 3 times as likely to be wearing sneakers as not wearing sneakers.
Ⓓ A randomly chosen person who is not wearing sneakers is 4 times as likely to be not wearing jeans as wearing jeans.

8.

	Enjoys Crosswords	Does not Enjoy Crosswords
Enjoys Sudoku	30	20
Does not Enjoy Sudoku	40	10

100 people were asked whether they enjoy crossword puzzles and whether they enjoy sudoku number puzzles. The results are shown in the table above.

What percent of all 100 people enjoy sudoku?

Ⓐ 20%
Ⓑ 30%
Ⓒ 50%
Ⓓ 60%

9.

	Enjoys Crosswords	Does not Enjoy Crosswords
Enjoys Sudoku	30	20
Does not Enjoy Sudoku	40	10

100 people were asked whether they enjoy crossword puzzles and whether they enjoy sudoku number puzzles. The results are shown in the table above.
Which of the following statements is NOT supported by the data?

Ⓐ A randomly chosen person who enjoys sudoku is more likely to enjoy crosswords than to not enjoy crosswords.
Ⓑ A randomly chosen person who does not enjoy sudoku is more likely to enjoy crosswords than to not enjoy crosswords.
Ⓒ A randomly chosen person who enjoys crosswords is more likely to enjoy sudoku than to not enjoy sudoku.
Ⓓ A randomly chosen person who does not enjoy crosswords is more likely to enjoy sudoku than to not enjoy sudoku.

10.

Preferred Sports

	Volleyball	Basketball	Softball
Boys	5	30	15
Girls	30	5	15

Out of those students who preferred volleyball, about what percent were girls?

Ⓐ 15%
Ⓑ 35%
Ⓒ 85%
Ⓓ 100%

11. **Using the data from the table below match the answers to the questions about the table. June surveyed the 7th and 8th grades to see which class they liked better, math or English. The results are shown in the two-way table below.**

	Math	English
7th grade	78	67
8th grade	86	45

	131	.60	145	.66
Total number of 7th graders surveyed	○	○	○	○
Total number of 8th graders surveyed	○	○	○	○
The relative frequency of 7th grade students that chose English to all students that chose English	○	○	○	○
The relative frequency of 8th grader students that chose Math to the total number of 8th graders	○	○	○	○

12. June surveyed the 7th and 8th grades to see which class they liked better, math or English. The results are shown in the two-way table below. Answer the question that follows.

	Math	English
7th grade	78	67
8th grade	86	45

The relative frequency of 7th grade students that chose math to all 7th grade students is ______

13. Sam surveyed his classmates to find out if they played a sport after school or practiced in the school band. Fourteen of his classmates played a sport. Of those 14, only 5 participated in the school band. Eight students played in the school band. There were ten students who did not play a sport nor participate in the school band. Fill in the blanks in the table given below.

	Plays a Sport	Doesn't Play a Sport	Total
Plays in Band	5		8
Does Not Play Band		10	
Total		13	27

End of Statistics and Probability

ANSWER KEY AND DETAILED EXPLANATION

Chapter 5: Statistics and Probability

Lesson 1: Interpreting Data Tables & Scatter Plots

Question No.	Answer	Detailed Explanation
1	B	By definition, a decreasing trend from left to right on a scatter plot indicates a negative association.
2	A	By definition, an increasing line from left to right on a scatter plot indicates a positive association.
3	C	The points in the third choice are nearly in a straight line.
4	A	As the History grade increases, so does the English grade. Thus, there is a positive association.
5	A	There does not appear to be any significant correlation between these two variables.
6	A	As the Science grade increases, so does the Math grade. This indicates a positive association.
7	A	In the first graph, a best fit line would be nearly a straight line increasing to the right.
8	A	$30\times\frac{100}{150} = 20\%$
9	B	Option (B) is the correct answer. 80 do not play an instrument and 50 of those do not play a sport. Among these 80 students, ratio of those who do no play a sport to those who play a sport is 50:30 = 5:3, It is not 2:1 as given in the option (B). All the other statements are supported by the data.
10	A	30+50=80, which is about half of 150.

Question No.	Answer	Detailed Explanation
11		If there is a positive association, as one item increases so does the other. If there is a negative association, as one increases the other decreases. The last one is no association. This happens if there is no pattern as to what is happening.
12	no association	There is no association between the apple and mango sales. You can see that they neither consistently go up or go down over the course of the 10 days.
13	negative|| none|| positive|| negative	The first graph has a negative association because as you move across the graph you can see that the points plotted continually decrease. The second graph has none because there is no pattern to the points. The third graph has a positive association because as you move across the graph you can see that the points continually increase. The last graph will also be a negative association because as you move across the graph you can see that the point plotted continually decrease.

Lesson 2: Scatter Plots, Line of Best Fit

Question No.	Answer	Detailed Explanation
1	D	This scatter plot does not represent a linear function. None of these is the correct choice.
2	C	The line of best fit would be a horizontal line. Constant linear is the correct choice.
3	D	The data does not represent a straight line. None of these is the correct choice.
4	A	The first scatter plot shows a positive slope representing a positive linear association.
5	B	The second scatter plot shows a negative slope representing a negative linear association.
6	D	The data on the fourth scatter plot cannot be represented linearly and, therefore, represents no association.
7	C	The third scatter plot represents a horizontal line or a constant association.
8	B	The data represents a straight line with a slope of positive 1. The second scatter plot is the correct choice.
9	A	The y-coordinate is a constant 4 representing a horizontal line at y=4. The first scatter plot is the correct choice.
10	C	Temperature decreases as the hours increase. Negative linear association is the correct choice.
11		To get the best fit line for the data, it won't necessarily go through every point, but will show the general trend of the data.
12	y=(-3/7) x+45/7	The best fit line is already drawn on the graph, but they want the equation to go with it. Using the 2 points they gave us we first need to find the slope. $\frac{6-8}{1-8} = -\frac{3}{7}$ Now that we know the slope, we can fill in an ordered Then, substitute an ordered pair into the equation y = (-3/7)x + b to solve for b. Using the point (1,6), solve 6 = $\frac{-3}{7}$(1)+b to find b = 45/7. Therefore, the equation of the line is y = $\frac{-3}{7}$x+ $\frac{45}{7}$

<table>
<tr><th>Question No.</th><th>Answer</th><th>Detailed Explanation</th></tr>
<tr><td>13</td><td>(1) 56
(2) 55, 80
(3) 81</td><td>
<table>
<tr><th></th><th>Doesn't Like Dramas</th><th>Likes Dramas</th><th>Total</th></tr>
<tr><td>Doesn't Like Comedies</td><td>41</td><td>56</td><td>97</td></tr>
<tr><td>Likes Comedies</td><td>55</td><td>25</td><td>80</td></tr>
<tr><td>Total</td><td>96</td><td>81</td><td>177</td></tr>
</table>
To find how many students like dramas but don't like comedies, subtract 41 from 97 to get 56. Next you know that 55 liked comedies but not dramas. To get the total number of students who liked comedies, just add 55 + 25 to get 80. Lastly to get the total of how many like dramas, take 56 + 25 to get 81.</td></tr>
</table>

Lesson 3: Analyzing Linear Scatter Plots

Question No.	Answer	Detailed Explanation
1	A	Visual inspection of the points and lines shows that the first scatter plot is the best choice because it shows a positive association, just like the original scatter plot does. As x increases, y increases.
2	A	The line in choice one passes through the center of the scatter plot with approximately the same number of stray points on each side.
3	D	The line in choice 4 is clearly a best fit line.
4	D	The line in choice 4 passes through nearly every point in the scatter plot.
5	C	The line in choice 3 has the strongest association with the points in the scatter plot.
6	C	The fifth point in choice 3 is an outlier because it is positioned farther away from the line of best fit than in the other scatter plots.
7	B	Choices 1, 3, and 4 have outlier points. Choice 2 does not.
8	B	We are told that the slope is 2. The slope is always the change on the vertical axis divided by the change on the horizontal axis; so a slope of 2 would be interpreted as 2/1 which represents 2 units on the vertical axis for every unit on the horizontal axis. Looking at the graph we see that the vertical axis is cm and the horizontal axis is days. Therefore a slope of 2=2/1 would represent 2 cm growth for each day.
9	A	Slope represents the (change on the vertical axis) / (the change on the horizontal axis). The vertical axis represents dollars and the horizontal axis represents miles; so a slope of 0.5 would be interpreted as 0.5 dollars per mile or 0.5 dollars/mile which is 50 cents for each mile traveled.
10	D	We are told that the graph relates the temperature in a school's parking lot, in degrees Fahrenheit, to the number of hours past noon. A slope of -3/1 = a temperature decrease of 3°F/hour.

<table>
<tr><th>Question No.</th><th>Answer</th><th>Detailed Explanation</th></tr>
<tr><td>11</td><td></td><td>
<table>
<tr><th></th><th>Linear</th><th>Negative Association</th><th>Line of Best Fit</th><th>Prediction Equation</th></tr>
<tr><td>A line on a graph showing the general direction that a group of points seem to be heading</td><td></td><td></td><td>●</td><td></td></tr>
<tr><td>A graph that is represented by a straight line</td><td>●</td><td></td><td></td><td></td></tr>
<tr><td>The equation of a line that can predict outcomes using given data</td><td></td><td></td><td></td><td>●</td></tr>
<tr><td>A correlation of points that is linear with a negative slope</td><td></td><td>●</td><td></td><td></td></tr>
</table>
The line that is on the graph that doesn't necessarily go through every point but represents the general trend of the graph would be the line of best fit. If it is a straight line, then it is a linear graph. The equation of the best fit line that helps predict outcomes is called the prediction equation. Lastly, if the trend of the scatter plot has a negative slope then it has a negative association.
</td></tr>
<tr><td>12</td><td>y=(1/2) x+3/4</td><td>
Using the 2 points they gave us we first need to find the slope. Now that we know the slope,

$$\frac{2-0}{2.5-(-1.5)} = \frac{1}{2}$$

We can substitute an ordered pair and solve for the y-intercept:

$$2 = \left(\frac{1}{2}\right)(2.5)+b$$

When you solve for b, you get 3/4. Putting these values together gives the equation y = (1/2)x + 3/4.
</td></tr>
</table>

Lesson 4: Relatable Data Frequency

Question No.	Answer	Detailed Explanation
1	C	Out of 25 people wearing sneakers, 15 were also wearing jeans. $\frac{15}{25}$ = 60% 60% is the correct choice.
2	D	Out of 20 people wearing jeans, 15 were also wearing sneakers. $\frac{15}{20}$ = 75% 75% is the correct choice.
3	B	Out of 25 people not wearing sneakers, 5 were wearing jeans. $\frac{5}{25} = \frac{1}{5}$ = 20% 20% is the correct answer.
4	B	Fertilized plants: $\frac{50}{200} = \frac{1}{4}$ died Unfertilized plants: $\frac{150}{600} = \frac{1}{4}$ died They died at the same rate is the correct choice.
5	D	There were a total of 9 windy days and 5 of them were also sunny. $\frac{5}{9}$=56% The correct choice is 50% of the windy days were also sunny.
6	B	Out of 25 people wearing sneakers, 10 were not wearing jeans. $\frac{10}{25} = \frac{2}{5}$ $\frac{2}{5}$ is the correct answer.
7	A	Out of 25 people wearing sneakers, 15 were wearing jeans and 10 were not. The statement, "A randomly chosen person who is wearing sneakers is equally as likely to be wearing jeans as not wearing jeans", is not supported by the data.
8	C	50 out of 100 people enjoy sudoku. The correct answer is 50%.

<table>
<tr><th>Question No.</th><th>Answer</th><th>Detailed Explanation</th></tr>
<tr><td>9</td><td>C</td><td>Out of 70 people who enjoy crosswords, 30 enjoy sudoku and 40 do not.
The statement, "A randomly chosen person who enjoys crosswords is more likely to enjoy sudoku than to not enjoy sudoku," is not supported by the data.</td></tr>
<tr><td>10</td><td>C</td><td>Out of 35 students who preferred volleyball, 30 were girls.
$\frac{30}{35} = 86\%$
85% is the correct choice.</td></tr>
<tr><td>11</td><td></td><td>
<table>
<tr><th></th><th>131</th><th>.60</th><th>145</th><th>.66</th></tr>
<tr><td>Total number of 7th graders surveyed</td><td></td><td></td><td>●</td><td></td></tr>
<tr><td>Total number of 8th graders surveyed</td><td>●</td><td></td><td></td><td></td></tr>
<tr><td>The relative frequency of 7th grade students that chose English to all students that chose English</td><td></td><td>●</td><td></td><td></td></tr>
<tr><td>The relative frequency of 8th grader students that chose Math to the total number of 8th graders</td><td></td><td></td><td></td><td>●</td></tr>
</table>

To find the total number of 7th graders you add 78 + 67 = 145. To get the total number of 8th graders surveyed add 86 + 45 = 131. To find the relative frequency of the 7th graders that chose English to the total number that chose English take $\frac{67}{67+45}$ = 0.60 (when rounded to the nearest hundredth). And lastly, to find the relative frequency of the 8th grade students that chose Math to the total number of 8th graders I took $\frac{86}{86+45}$ = 0.66 (when rounded to the nearest hundredth).</td></tr>
</table>

<table>
<tr><th>Question No.</th><th>Answer</th><th>Detailed Explanation</th></tr>
<tr><td>12</td><td>0.54</td><td>To get the relative frequency you will take the number of 7th grade students that chose math and divide it by the total number of 7th graders. In this case you will take 78 and divide it by 145. When you do this, you get 0.54 (when rounded to the nearest hundredth).</td></tr>
<tr><td>13</td><td>(1) 3
(2) 9; 19
(3) 14</td><td>
<table>
<tr><th></th><th>Plays a Sport</th><th>Doesn't Play a Sport</th><th>Total</th></tr>
<tr><td>Plays in Band</td><td>5</td><td>3</td><td>8</td></tr>
<tr><td>Does Not Play Band</td><td>9</td><td>10</td><td>19</td></tr>
<tr><td>Total</td><td>14</td><td>13</td><td>27</td></tr>
</table>

Total Number of students who play band is 8 and we know who play band and also play a sport is 5. Hence, those who don't play a sport but play band is 8 - 5 = 3
We know the total number of students who play sport is 14. Hence third row, first column will be 14. Since this total is 14, Not in band but plays sports will be 14 - 5 = 9. Hence, total of that row will be 9 + 10 = 19.</td></tr>
</table>

Notes

Additional Information

Test Taking Tips

1) **The day before the test,** make sure you get a good night's sleep.

2) **On the day of the test,** be sure to eat a good hearty breakfast! Also, be sure to arrive at school on time.

3) **During the test:**

- **Read every question carefully.**
    - Do not spend too much time on any one question. Work steadily through all questions in the section.
    - Attempt all of the questions even if you are not sure of some answers.
    - If you run into a difficult question, eliminate as many choices as you can and then pick the best one from the remaining choices. Intelligent guessing will help you increase your score.
    - Also, mark the question so that if you have extra time, you can return to it after you reach the end of the section.
    - Some questions may refer to a graph, chart, or other kind of picture. Carefully review the graphic before answering the question.
    - Be sure to include explanations for your written responses and show all work.
- **While Answering Multiple-Choice (EBSR) questions.**
    - Select the bubble corresponding to your answer choice.
    - Read **all** of the answer choices, even if think you have found the correct answer.
- **While Answering TECR questions.**
    - Read the directions of each question. Some might ask you to drag something, others to select, and still others to highlight. Follow all instructions of the question (or questions if it is in multiple parts)

Frequently Asked Questions(FAQs)

For more information on the Assessment, visit
www.lumoslearning.com/a/tcap-faqs
OR Scan the **QR Code**

What if I buy more than one Lumos tedBook?

Step 1 → **Visit the link given below and login to your parent/teacher account**

www.lumoslearning.com

Step 2 → Open the **My Account** menu, go to **My Subscriptions**, and select **My tedBooks**. Enter the book access code shown on the first page of the book and submit.

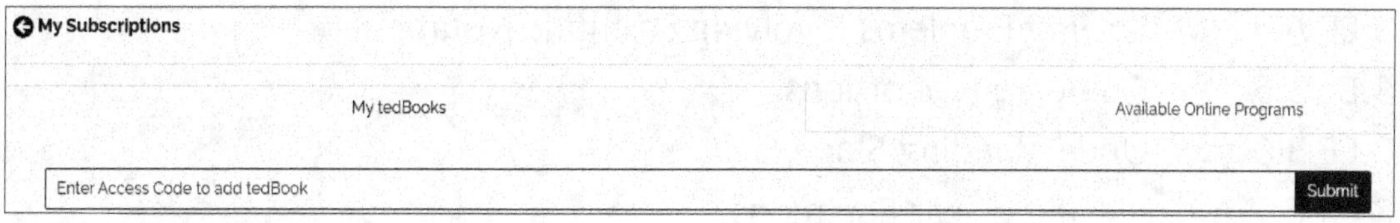

Step 3 → **Add the new book**

To add the new book for a registered student, choose the '**Student**' button and click on submit.

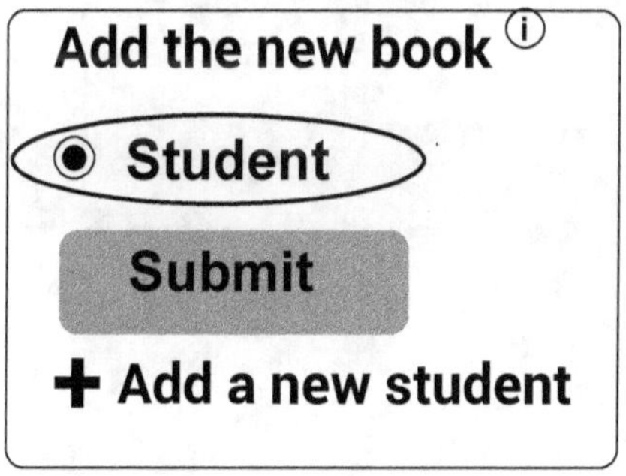

To add the new book for a new student, choose the '**Add New Student**' button and complete the student registration.

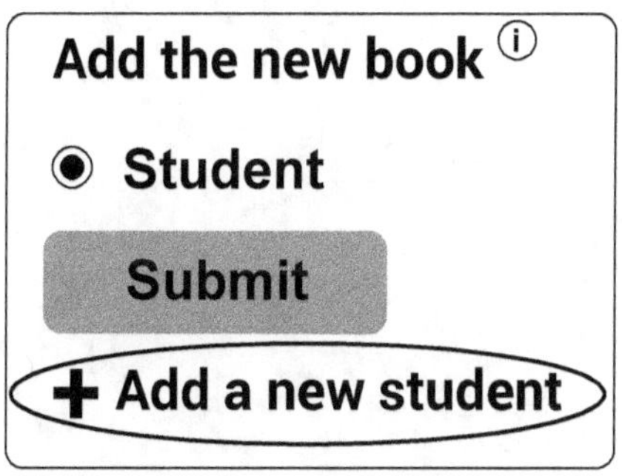

Progress Chart

Standard	Lesson	Score	Date of Completion
TAS			
8.NS.A.1	Rational vs. Irrational Numbers		
8.NS.A.2	Approximating Irrational Numbers		
8.EE.A.1	Properties of Exponents		
8.EE.A.2	Square & Cube Roots		
8.EE.A.3	Scientific Notations		
8.EE.A.4	Solving Problems Involving Scientific Notation		
8.EE.B.5	Compare Proportions)		
8.EE.B.6	Understanding Slope		
8.EE.C.7.A	Solving Linear Equations		
8.EE.C.7.B	Solve Linear Equations with Rational Numbers		
8.EE.C.8.A	Solutions to Systems of Equations		
8.EE.C.8.B	Solving Systems of Equations		
8.EE.C.8.C	Systems of Equations in Real-World Problems		
8.F.A.1	Functions		
8.F.A.2	Comparing Functions		
8.F.A.3	Linear Functions		
8.F.B.4	Linear Function Models		
8.F.B.5	Analyzing Functions		

Standard	Lesson	Score	Date of Completion
TAS			
8.G.A.1	Transformations of Points & Lines		
8.G.A.1.B	Transformations of Angles		
8.G.A.1.C	Transformations of Parallel Lines		
8.G.A.2	Transformations of Congruency		
8.G.A.3	Analyzing Transformations		
8.G.A.4	Transformations & Similarity		
8.G.A.5	Interior & Exterior Angles in Geometric Figures		
8.G.B.6	Verifying the Pythagorean Theorem		
8.G.B.7	Pythagorean Theorem in Real-World Problems		
8.G.B.8	Pythagorean Theorem & Coordinate System		
8.G.C.9	Finding Volume: Cone, Cylinder, & Sphere		
8.SP.A.1	Interpreting Data Tables & Scatter Plots		
8.SP.A.2	Scatter Plots, Line of Best Fit		
8.SP.A.3	Analyzing Linear Scatter Plots		
8.SP.A.4	Relatable Data Frequency		

www.ingramcontent.com/pod-product-compliance
Lightning Source LLC
LaVergne TN
LVHW080045210926
848481LV00011B/3283